Benjamin Rush, A. H. Flanders

The Family Physician

Consumptives Guide to Health and Lady's Medical Companion

Benjamin Rush, A. H. Flanders

The Family Physician
Consumptives Guide to Health and Lady's Medical Companion

ISBN/EAN: 9783337121075

Printed in Europe, USA, Canada, Australia, Japan

Cover: Foto ©berggeist007 / pixelio.de

More available books at **www.hansebooks.com**

...CIAN,

...NSUMPTIVE'S

HEALT...

AND

LADY'S MEDICAL C...

COMPILED CHIEFLY FROM...
BENJAMIN RUSH, M....

PROFESSOR OF CHEMISTRY, THEORY AND...
OF MEDICINE IN THE UNIVERSITY...
U. S. SURGEON GENERA...

EDITED BY...
A. H. FLANDERS, A...

LATE PROFESSOR OF CHEMISTRY, CH...
COLOGY, IN THE HOM. MEDICAL COLL...
GRADUATE OF UNION COLLEGE, ...
BER OF THE RHODE ISLAND...
ETY; MEMBER OF TH...
MANNIAN INSTIT...
CORRESPONDING SECRETARY OF THE IM...

LOWELL:
PUBLISHED BY THE EDITOR.

CONSULTATIONS FREE.

PROF. FLANDERS, the editor of this book, makes no charge for consultation, or advice, whether personally or by letter. His residence, and office, are in Wentworth's Building, Lowell, Mass.

For the benefit of those who cannot visit the city, he has so arranged his practice and prepared his remedies, that he can treat them successfully and satisfactorily at a distance; although he prefers that the patient should give him the opportunity of a personal examination, whenever it is possible, by calling at his office. But when this cannot be done conveniently, and invalids can only consult by letter, a printed list of questions will be fowarded, by which all the facts necessary to any case will be ascertained. A candid opinion will be promptly returned.

Remedies may be sent by express to all parts of the country. His treatment, conducted upon this plan, has been attended by the most gratifying results, and its benefits have been extended to thousands who could not leave their distant homes, and must otherwise have failed of relief. The remedies he employs are carefully prepared under his own hand.

His charges are moderate, and the poor should not hesitate to apply. *The full address is given at page* 44.

INTRODUCTION.

This book lays no claim to originality ; but, based upon the works of the great Dr. Rush, its object is simply utility. With this in view, no trouble has been spared to adapt it to the present state of medical science, and render it perfect in its way.

The Laws of Health are considered in relation to the prevalent habits, and mode of life of our people ; and in the hope that they may become more generally known, and observed, than at present. The directions under Bathing and Water Treatment, are such as long personal experience has proved to be most essential for the preservation and restoration of health.

Those receiving this work as the Consumptive's Guide to Health, are advised to read over all the matter commencing with page 6, until the article on Consumption is finished.

It is to be hoped that the fastidious will not take offence at any thing contained herein ; remembering that " to the pure, all things are pure ;" and that the physician of the body, like the Good Physician, must be ready to assist all classes of the unfortunate.

DIRECTIONS TO THE READER.

Almost any topic upon which medical information is usually wanted may be found by looking in the alphabetical index at the end of the book.

The following topics may be found more especially interesting. The Laws of Health, page 8; Scrofula, page 40; Rush's Sarsaparilla and Iron, page 47; Rush's Lung Balm, page 54; Cures of Consumption, page 58; Consumption, page 64; Medical Humbugs, page 143; Accidents, page 149; Poisons, page 152; Fevers, page 153; Dyspepsia, page 164; Rheumatism, page 168; Diphtheria, page 177; Cancer, page 183; Asthma, page 188; Diseases of the Heart, page 189; Diseases of the Skin, page 190; Female Complaints, 210; Rush's Monthly Remedy, page 231; The Laws of Maternity, or How to prevent an increase of Family, page 232; Syphilis, page 235; Gonorrhœa, page 240; Self-Abuse, page 241; Receipts, page 246.

The receipt for Chromate Soap, housekeepers will find very valuable, page 246.

The Editor's Tour in Europe.

In answer to the inquiries of numerous friends, the editor will take this opportunity to make a brief response, in regard to his recent visit to the principal countries of the Old World.

His journey was one of observation, and improvement ; and he embraced the opportunity to make himself acquainted with the different modes of treatment, and accumulated experience of British and Continental physicians. For nearly a year he devoted his time to visiting the European hospitals, availing himself of the researches and knowledge of the most eminent surgeons and physicians in Europe ; such as Sir B. Brodie, Curling, Civiale, Ricord, and Guerin ; and thinking medical chemistry of the utmost importance in securing absolute purity for all medical preparations, especially in vegetable chemistry, he made himself thoroughly acquainted with the best Laboratories in England and on the Continent ; especially those of Liebig, Regnault, and Kane. His tour thus extended through Great Britain, France, Italy, Germany and Switzerland ; visiting also the principal hospitals of London, Paris, Edinburg, Rome, Milan, Venice, and Vienna. On his return he found himself amply repaid by the additional knowledge acquired in the treatment of various diseases, especially of the more delicate and difficult diseases that devastate the human race ; such as Consumption and all disorders of the Pulmonary tissues ; and those of the Urinary, Nervous, and Sexual functions.

It is well known by many physicians that the means commonly used in the curing of these diseases have baffled all their efforts; and none can more readily testify to the truth of this assertion than the poor patient himself, who has proved it by bitter experience.

There is, undoubtedly, an aptitude in particular systems to particular diseases, or a *something* causing that disease to assume an unusual inveteracy. It was the good fortune of the editor while residing at Paris, to witness in the hospitals every form of disease in the pulmonary and nervous systems, and the infallible means adopted for their removal and cure

He found the facilities for medical improvement, at Paris to be far superior to those of any other European capital, equally notable in Diseases of the Skin, and the more difficult, and obscure Female Complaints. The principal surgeons, and physicians, attached to the various medical institutions there, are not only polite, but quite painstaking, in giving every needed information to physicians from abroad, and especially to Americans. The authorities of the Empire, too, are no less liberal in according every facility for attending the great government hospitals devoted to special diseases

To give any detailed account of the medical facilities of Paris and the Continent, would fill a volume; or of their artistic and literary advantages;—it will not be attempted here; but numerous friends may rest assured that however great such facilities, and advantages, may have been found; old Europe had no permanent attractions, for Young America.

THE LAWS OF HEALTH.
HYGIENE. DIET.

Preservation of Health.

Pure Air, and plenty of it, that is to say good ventilation, is to be regarded as a matter of the first importance in preserving, or restoring health. Buildings public, as well as private, are now nearly always constructed with reference to better ventilation than formerly; especially in our cities, and larger towns: but in the rural districts, in fact, in all quarters, there are still great faults in this respect. Not only in dwellings, but in churches, school-rooms and all places of public meetings, is ventilation, at times, neglected; and the breathing of impure air thereby is a most fruitful source of disease.

Proper ventilation is not practicable, and consistent with health without due regard to temperature. A thermometer is therefore an essential instrument in every well regulated room, which is to be kept warm; and especially for a sick room. The proper temperature for comfort and health is between 65 and 70 degrees, when fires are kept; and *in all rooms warmed by close stoves, except in the very coldest weather, a window should be kept constantly let down at the top, if only an inch or two;* and **this** is especially true of new, tight, and mod-

ern-built houses. A very good way to secure ventilation in very cold weather, is to have a register in-
serted in the chimney, and opening out of the room
to be ventilated. This may usually be done by
putting it over the fire-place. This is also a good
way in sleeping rooms when persons cannot make
up their minds to sleep with a window partly or
quite open. Sleeping rooms, however, are best
ventilated by opening a window wide open, and
allowing the pure air of all-out-doors to enter
freely. This may be practiced with perfect safety,
even all winter, and in stormy weather, if there
are protecting blinds to keep out rain or snow, with
the greatest advantage to sound sleep and health.
The only exceptions are to zero weather, and very
windy weather, when enough air for ventilation
will enter if a window is open a half inch, or so.

There is a prevalent idea that sleeping in a
room with a fire is unhealthy, and this is no doubt
correct if the ventilation is bad, as it usually is
when no window is left open ; because the fire
helps consume the vital air. But if a window is
left open a fire is by no means unhealthy. Some
fear taking cold from leaving a window open : —
there will be no danger for persons who take a
daily morning bath; and others must accustom
themselves by degrees ; that is by leaving a window
a little farther open every night, or every week,
until the habit is formed. Young persons especially should form this habit. The objection is
sometimes urged that night air is unhealthy, and
injurious. We are certain that this is only popular prejudice ; there is no chemical difference between air during the night and during the day,
except that in the night it may perhaps contain a
little more dampness; but if that is the case it is

rather an advantage; since it is one of the evils of our climate that the atmosphere is *too dry.* We repeat it that it is only a groundless prejudice that night air is injurious to health. Sleep then with a window open at all times of year, and promote, and preserve your health by so doing.

In general there is too little sound sleep. Eight hours is not too much for the average of persons; and how many there are who shorten their lives by trying to get along with five or six. No clothing should be worn by night, that is worn in the day time, for the reason that by making an entire change every night, the body clothing, or linen,' becomes thoroughly ventilated, and the most of the perspiration absorbed during the day evaporated. It is a good idea to place the immediate body linen, when it is to be resumed the next morning near the open window. Sleeping a short time just *before* dinner, is as good or better than a long nap *after.*

Feather beds, when very soft, are justly condemned as unhealthy. But one of the hard kind, made of less costly feathers, can hardly be worse than a hair mattress, which does not admit of stirring up, or as thorough ventilation as does a feather bed. We are inclined to think that the indiscriminate abuse of feathers is hardly fair; and do not believe that feathers are more retentive of perspiration, or odors, than curled hair. The old fashioned straw-bed is rather cold for winter, but is perhaps the healthiest of all in summer. This question, however, is of very little importance when compared with that of ventilation by open windows; or of daily bathing.

Most persons, and all sensitive ones, will find on trial that they will sleep better with the head

towards the North. The reason for which is, that
there is a force similar to magnetism, which Reichenbach has called the Odic force, which makes this
a great requisite for sound sleep with very many
persons.   Moonlight, or the direct rays of the
moon, in a sleeping room will frequently for the
same reason, make sensitive persons restless, or
even sleep-walkers.   Persons desirous of a further understanding of this subject should read the
Baron Reichenbach's " Dynamics of Magnetism."
We only state what are now established facts.

In this connection it may be stated that it is a very
good habit to drink a tumbler of pure water on going
to bed, in compensation for the loss by insensible
perspiration, and evaporation from the lungs, during the night.   It is certainly conducive to sound
sleep, and beneficial to persons inclined to dyspepsia, or constipation.

In regard to bed clothing it should be stated that
cotton-wadded quilts, or comforts, are objectionable on account of retaining the exhalations from
the body:—blankets are preferable, and can be
washed if requisite.

Bathing a Means of Preserving Health.

*Daily bathing is the most important means of all,
for preserving health.*   This is not entirely on account of personal cleanliness, which is a duty we
owe to society and ourselves ; but for the physiological reason, among others, that the daily application of cold water to the entire surface of the body
hardens and fortifies the entire system against any
and all disturbing causes ;—such as taking cold,
susceptibility to contagion, and all other causes
productive of disease.   It also keeps the system in

a constant state of renewal; because oathing causes a greater waste of the solids than would otherwise take place, but which is more than made up, by the increased appetite and more perfect digestion, by which we get new and perfect blood, and from that, a renewal of the solids, in a condition resembling that of youth. We believe it to be a fact that the average length of human life would be increased fifteen years if everybody could, and would, bathe at least once daily. Set it down, then, gentle reader, as an established fact that a daily bath of some kind will promote your health, and comfort, and probably lengthen your life.

The question is then asked what kind of a bath is best. We answer from long personal experience

The Sponge Bath.

The Sponge Bath taken on rising in the morning, is on many accounts the most effectual, and convenient bath which is avilable for everybody. A large sponge or coarse towel, or two, are all that are requisite save a basin of cold soft water. Use it in this way. Immediately on rising, the night clothing being removed, take a spongeful of water, and squeeze it out on the back of the neck, and by stooping a little let the water flow all over the body; repeat this over the chest and continue to take up rapidly one spongeful after another until one or two quarts have been briskly applied. Then with a large coarse towel briskly rub dry the whole surface.

If the hair is not too abundant wash the entire head and face with the hands, before you com-

mence the bath with the sponge. A coarse towel
may be used instead of a sponge. The very best
kind of towel for drying is a Turkish bathing towel. This entire bath, and wiping dry should occupy just *three minutes* and no more ; and you will
find it the best hour's work that you can do in the
day. The water which is used in taking this bath
may be received by standing in a common wash-
tub. A skilful person, however, can take it without spilling enough water on a carpet even, to do
any harm ; but it is better to use more water.
The body should be briskly rubbed with the coarse
towel at least *one minute*; which will be enough
in practiced hands. For the first few times the
entire bath may take up four or five minutes.
Persons unaccustomed to bathing need not bathe
the *entire* surface of the body for a beginning ;
but may commence with the face and chest only,
and gradually increase.

Perhaps you ask ;—Do you advise this bath to
be taken every morning, even if it freezes in the
sleeping room. We reply, yes, even if you break
the ice to get at the water. We have often done
so, and expect to do so many times again. If however, you commence bathing in winter, at first you
must use the water at about 65 degrees, until you
become accustomed to the oddity of it. We venture to say that after you have taken such a daily
bath as this for the period of one month you will
not willingly go back to the old greasy, sticky,
sleepy feeling, of an unwashed skin. Daily bathing renders the wearing of flannel next the skin;
the sleeping with hermetically sealed windows, for
fear of taking cold ; and in short, the whole " cod-
dling " process, wholly unnecessary. The editor
of this book was brought up in that way ; but

thanks to the progress of ideas " has come up higher." Perhaps you ask is not the shower-bath better than the sponge-bath. We answer no. It is a much harsher bath, but no better. Perhaps in summer it is a more pleasant bath, for daily bathing. We should prefer, if it were always accessible, a full bath, by immersing the whole person, daily on rising, and occupying the same length of time;— namely *three minutes*; but unfortunately very few persons enjoy such facilities.

We repeat it again, *the morning sponge-bath, taken daily on rising, as previously directed, is the best, and most powerful means of preserving health,* and available under all circumstances. After taking any kind of bath, warmth, and reaction should be secured by exercise; or if necessary, by going to a fire.

Food and Diet.

This is a subject both important and interesting ; and one we are sorry to say very little understood. We must remember that what we eat must be immediately digested, and converted into a substance called chyle, which enters immediately into the blood, and mixes with it; and that this is all a process of only a few hours. How important then to eat only such kinds of food as are proper to become a part of our bodies.

A point of very great importance is that of the *quantity* of food to be taken. In this country most persons eat too much, and especially too much animal food ; and excess in eating is considerably promoted by the almost universal use of fine flour which is a form of food too concentrated for health. Wheat contains all the chemical elements which

are necessary to invigorate the body : but some of
its most essential parts are contained in the cuticle
or bran, and the colored part of the kernel which
is next to the bran.   The small portion of iron,
which is essential to the constitution of healthy blood,
is there situated ; and the bran itself is necessary
to most persons to keep the liver and bowels sufficiently active.   If the use of fine flour could be
entirely abandoned, and the use of meal, made by
grinding up the entire wheat, be adopted instead ;
the health of the community would be very much
promoted.   The entire dyspeptic train of symptoms
is caused in no small degree by this excessive use
of fine flour, which has become a great and crying
dietetic evil in our land.   This is made worse, too,
by the almost universal use of unleavened bread,
in the form of biscuits, raised with saleratus, or
soda, which are still more injurious by reason of
loading the blood with pernicious mineral matter,
produced by the combination of the soda with the
cream of tartar, which forms a tartrate of soda.
Saleratus is just as bad.   How much better the
raised loaf of wheat meal ; or even the old-fashioned
loaf of brown bread, though a little less nutritious.

An excellent way of cooking wheat is to boil it
whole, or unground, as you would rice ; eaten with
milk or syrup, it forms a very nice and nutritious
dish, which a person could almost live on the entire year.   The only objection made by housekeepers is that it requires very long boiling ; but
this may be very much obviated by scalding the
wheat over night ; or by using it coarsely ground
like what is sometimes called hominy.   Rye and
Indian Corn should be eaten in the same way, but
are less objectionable even when bolted, than fine
wheat flour, for the reason that they are less con-

centrated forms of food, Valuable rules may be found among the "Receipts" for unbolted wheat and coarse bread. It will be proper to add that *lard* for shortening is a most unhealthy addition. All kinds of shortening are injurious; suet, and sweet butter the least so; but *lard* is the worst of all.

We have already stated that too much animal food is eaten. As a rule, once a day for it, is quite enough, even for laboring men; a pound of wheat at one fourth the cost contains as much substantial nourishment as a pound of meat; and it will be found so on trying the experiment. But if meat eating is carried to an excess in this country what shall we say of pork-eating that most unhealthy of all ways of taking food. It will be mentioned under Consumption and Scrofula that pork-eating is one of the most fruitful causes of those diseases; and the reasons there given; but it is not less a cause of many cases of Dyspepsia, Liver-Complaint, Constipation, and even Bilious Fevers. It is the most indigestible of all kinds of food, requiring according to the experiments of Beaumont and others, no less than six hours for its perfect digestion, while wheat and most vegetables require but an hour and a half, and beef but three hours. Pork is especially pernicious to all sedentary persons, and those who do not live much in the open air. To any person of a reflecting turn of mind it must be a disgusting idea that the flesh of a swine should be taken as food, and immediately become a part of our bodies; and thus the seeds of Scrofula, with which all swine are more or less contaminated, be habitually introduced into the system. The objection of possible disease is also good against all kinds of meats; since we can rarely tell by its appearance, that the animal

2

when killed was "perfectly healthy.   Scrofula,
however, is not usual in other animals, but as appears from the traditions of Eastern nations originated in the hog.   However that may be, the fact
is now certain.   It is frequently answered to this,
by farmers and others who have always been great
pork-eaters, and still are reasonably healthy persons, that their own experience proves the contrary.
To which we answer that the experience of many
persons who drink rum, and use tobacco all their
lives, might prove the same thing :— viz, that rum
and tobacco are healthy articles.   Take a dozen
of your acquaintances who have Scrofula, or have
died of Consumption, and you will find that every
one has been a pork-eater.   The Jews eat no
pork, and are very free from these diseases.   Being
much in the open air is a strong protection against
all morbid influences ; and so it is in the case of
farmers against pork-eating ; yet nearly one third
of them die of Consumption.

Most lean meats are too much cooked, and this
is especially true of the process of frying, and
frying in lard is the worst of all.   Boiling is preferable, and should be done quickly, and over a
hot fire.   Many receipts, and directions for cooking in accordance with the laws of health will be
found under " Receipts."

Coffee and Tea are both injurious articles ; and it
is perhaps fortunate that their present high price is
leading to their disuse.   According to experiments
made by the editor, tea, and coffee. when highly
concentrated proved active poisons, in doses of ten
drops, when given to cats and rabbits ; destroying
life in a few minutes.

Coffee produces piles, and other dyspeptic
troubles ; and tea may cause nervous disorders ;

but in the form of weak black tea is not as unhealthy as coffee. Keep your children from using
these articles, even if you cannot abandon the use
of them yourselves. Their use, as well as that of
tobacco, or alcohol, is an acquired habit. A young
child will instantly refuse them all, when taken
into the mouth for the first time, as being contrary
to his sense of taste, and instinctive perception.

The use of Salt.

The use of Salt is generally an abuse; most
persons use too much, and many suffer from dyspeptic and other chronic complaints caused by it,
without knowing it. An eminent writer thinks
that the use of salt is largely concerned in the
production of Cancers and diseases of the glands;
and that it is directly conducive to Scrofulous,
pulmonary, and skin diseases, and disorders of
the mucous membranes; also that it predispo-
ses to disease, and aggravates it when produced.
The use of salted food solely, certainly causes scurvy; and there is no doubt that the system does
not *require* any more than a minute quantity of
salt. Enough is taken in the butter we eat, to
supply the system.

Condiments generally.

The Condiments mustard, cayenne pepper, horse-
radish, stimulating sauces, garlic and ginger are all
injurious; and far from assisting digestion according to Dr. Beaumont's experiments actually retard
it.

Alcohol and Tobacco.

The subject of temperance has exhausted that of the abuse, or use as a beverage, of alcoholic stimulants. Few deny that they are injurious; but very many, and even clergymen, who would consider it a sin, and a shame to use ardent spirits daily; openly, or covertly use tobacco. It is a very powerful narcotic and acrid poison; the oil of which will destroy human life in less than ten minutes, or as quick as Prussic Acid. Fatal effects have frequently resulted from the medicinal use of tobacco. As the result of its habitual use the main injury is usually done to the brain and nervous system, and to the lungs. Loss of memory; general weakness of the nervous system, producing trembling of the hands, palpitation, and the like; and softning of the teeth, often follow its use. For its effects on the lungs, see " Causes of Consumption."

Every young man who has not yet used tobacco, is earnestly advised never to form the habit. Few have the moral courage to abandon it.

Things that everybody may eat.

We have mentioned several things which it is better not to eat; some are allowable, but not to be paticularly commended; such as sweet butter, fresh eggs lightly boiled, fresh fish, shell fish, fresh beef, mutton, veal, and lamb, poultry, and game, a little vinegar, and unspiced pickled cucumbers. The more commendable articles are *Cereals*, Vegetables, and Fruits, which should constitute the bulk of our living. We have already spoken of wheat, rye, and Indian corn, to which may be added oat meal, as useful cereal productions. Among vegetables, potatoes are the most nutritious, and healthy; but require care in cooking; beans, and peas, are perhaps the next in order of nutrition, and utility, and for health should not be cooked with pork, but fat beef instead. Beets, carrots, and parsnips will do for those who like them; asparagus is allowable in its season, so of radishes and cucumbers; lettuce is unhealthy. Most persons eat two little fruit, and they are the same who eat too much meat. Of all the fruits the apple is the most valuable, and is a truly healthful article of diet. It may be eaten raw, or variously cooked. Baked sweet apples and milk are very rich, and nutritious.

Pears, peaches, grapes, plums, currants and the smaller fruits, may all be eaten in their season; or moderately when preserved. The plan of preserving them when fresh, in sealed jars without sugar, is best. Tomatoes are well worth preserving in this way.

We shall here annex a list of articles of food which are commendable, and allowable; and another of those which are unhealthy for most or all persons.

Commendable, or Allowable Food.

Soup or Broth from the lean of beef, veal, and mutton: to which may be added well boiled rice, barley, wheat, sago, tapioca, or maccaroni, seasoned merely with a little salt, if desired.

Meats, beef, mutton, poultry, and game, plainly cooked, and *not* fried in lard. *Fish.* Oysters and most kinds of fresh fish. Salt fish, *very* salt as it usually is, is not healthy except as a relish.

Vegetables. Potatoes, tomatoes, beans, peas, spinach, beets, turnips, carrots, parsnips, cauliflower; and for persons in health, cucumbers, and asparagus; cucumbers, however, do not agree with everybody.

Bread. All kinds of raised bread; but best of all that made of unbolted wheat meal; or brown bread, made of Indian meal and wheat or rye meal mixed. Biscuit free from soda and saleratus. *Eggs* lightly cooked.

Light Puddings; such as *boiled wheat or wheat meal*, hominy or hasty pudding, rye hasty pudding, indian pudding, sago, rice, bread. Simple cakes composed of flour or meal, eggs, sugar, and a little good butter. Apple puddings.

Fruit. Baked, stewed, or preserved *apples* or pears. Raw apples. The smaller fruits ripe, or preserved. Honey, syrups, and rich preserves in moderation. Preserves with sugar unspiced. Home made citron preserved. Raisins and grapes. *Drinks.* Water, milk, weak black tea, cocoa, chocolate, (unspiced), and for the sick, arrow-root, or gruel made thin, toast-water, barley water, milk and water, sugar and water, rice water, *wheat jelly water.*

Food prohibited for the Sick and usually unhealthy for all.

Soups. Turtle, mock-turtle, and all kinds of rich and seasoned soups.

Meats. Pork, bacon, duck, goose, sausages, kidney, liver, and all fat and salted meats.

Vegetables. Lettuce, celery, onions, artichokes, parsley, horse-radish, beets, thyme, garlic, salads, and pickles greened with copper.

Pastry of all kinds, but especially mince pie, and fried and boiled kinds. Anything fried in lard. Lard for shortening.

Spices, Aromatics, and Artificial Sauces of all kinds, and mustard:—also vinegar, and pickles when taking medicine. *Cheese.*

Fruit. Oily nuts, Confectionery generally.

Drinks. Coffee, green tea, all malt and spirituous liquors.

General Rules.

The following rules may be deduced from the foregoing remarks, founded on the observations of the best physiologists.

I. Bulk is nearly as necessary to food, as the nutritious principle. The coarse parts of food, therefore, should be eaten with the fine. Too highly nutritive diet is nearly as fatal to life and health as that which is insufficient in nourishment.

II. The more plain and simple the preparation of food, and the less of seasoning of any kind, the better for health. Stimulating condiments, such as cayenne pepper, mustard, &c., instead of being of any use, are actually injurious to the healthy stomach. And though they may assist the action of an enfeebled stomach for a time, their continued use never fails to produce its indirect debility. They affect it as achohol or other stimulants do ;— the present relief is at the expense of future suffering.

III. Thorough chewing and slow swallowing are of great importance.

IV. A due *quantity* of food is of the utmost importance. There is no subject of dietetic economy about which people are so much in error as that which relates to *quantity.* *Dyspepsia is more often the effect of overeating and overdrinking than any other cause.*

V. Solid food, if properly chewed is more easy of digestion than soups and broths.

VI. Pork, fat meat, and all oily substances, being always of hard digestion, tending to derangement of the stomach, are better omitted.

VII. Alcoholic liquors of every form, the various stimulating condiments, as mustard, cayenne pepper, spice, &c., tea, coffee, and narcotics of every kind, all tend to debility, derangement and disease of the stomach, and, through it, of the whole system.

VII. *Simple pure water is the only fluid necessary for drink, or for the wants of the system.* The artificial drinks are *all* more or less injurious. *Tea* and *coffee,* the common beverages of all classes of people, have a tendency to debilitate the digestive organs. Let any one who is in the habit of drinking either of these articles in a weak decoction, take two or three cups, made very strong, and he will soon be aware of their injurious tendency; and this is only an *addition* to the strength of the narcotic which he is in the constant habit of using.

Clothing.

Ours is an *extreme* climate, in which much more clothing is required in winter, and less in summer, than in most other parts of the temperate zones. It therefore becomes an interesting question, how much clothing, and what kind, shall be worn. Most persons wear too much by reason of excessive precaution. This has led to the wearing of woolen flannel next the skin; a useful thing

for persons who do not bathe daily, or sleep with
an open window; but quite unnecessary for those
who do. *Priessnitz*, the originator of the Water-
Cure practice required his patients to lay aside
their flannels;—many would object;—to whom he
would reply that flannel worn next the skin rendered people delicate and less able to contend a-
gainst atmospheric changes. He then tells them to
wear it *over* the linen, until they become accustomed to cold water bathing, when they can leave it off,
and will not miss it. He directs exercise after
each bath until slight perspiration begins, when
there will be no fear of taking cold.

Clothing should nowhere injuriously press upon
the body. It should be well proportioned over all
parts, and all sudden changes avoided. In the articles on " Consumption," and " The Prevention
of Consumption," may be found some useful hints
on *Exercise, Pure Air, and Clothing.*

Are the above Rules practiable ?

Substantially the above question is often asked.
We answer decidedly, yes. We have put them
personally in practice, as far as circumstances
would admit, for many years; and have seen them
put in practice by many others. They are not
theoretical merely, but eminently practical. *We
practice what we preach.* In substantiation of reformation in the Laws of health, read the following abstract from an extensive experiment made
at the Orphan Asylum of Albany, N. Y.

This institution was established in 1830. Shortly
after it contained 70 children, and subsequently
many more. For the first three years the diet
consisted of fine bread, rice, Indian puddings, po-

tatoes and other vegetables and fruit with milk; to which was added flesh, or flesh soup once a day. Moderate attention was paid to bathing and cleanliness, and to clothing, air, and exercise.

Bathing, however, was performed in a perfect manner only once in three weeks. Many were received in poor health, and not a few continued sickly.

In the fall of 1833 the diet and regimen of the inmates were materially changed. *Daily ablution of the whole body*, in the use of the cold shower or sponge bath, or, in cases of special disease, the tepid bath, was one of the first steps taken; then the fine bread was laid aside for that made of unbolted wheat meal, and soon after flesh, and flesh soups were wholly banished; and thus they continued to advance, till in about three months more they had come fully upon the vegetable system, and had adopted reformed habits in regard to *sleeping, air clothing, exercise*, &c. They continued on this course till August, 1836, when the results were as follow:— During the first three years in which the old system was followed, from four to six children were continually on the sick list, and sometimes more. A physician was needed once, twice, or three times a week, uniformly, and deaths were frequent. During this whole period there were between thirty and forty deaths. After the new system was fairly adopted, the nursery was entirely vacated, and the services of the nurse and physician no longer needed, and for more than two years no case of sickness or death took place. In the succeeding twelve months there were three deaths, but they were new inmates, and were diseased when admitted. Their condition afterwards continued to improve, not only as to health, but also in mental cheerfulness, contentment, activity, vivacity, and happiness. Statements from boarding schools for children in Germany, conducted on a similar state of facts, when the editor was there.

Water has been spoken of as the best and only beverage for the health. It should be *pure* water, and that is always *soft* water. As a rule water which is too hard to wash with, is not fit to drink; and this is always the case in lime-stone sections. When the hardness, or mineral matter, consists of clay, it is less injurious. Cisterns and filters must often be depended upon for pure water.

Water Treatment.

The use of daily bathing for preserving health has already been noticed; but water is not less valuable as a means of curing diseases: or at least driving them out of the system. The limits of this work will not allow of a fair exposition of the Hydropathic system; but the reader may rest assured that its processes are exceedingly valuable. The water-cure has been called the luxury of the rich; but both rich and poor may enjoy its advantages if they will only lay aside those prejudices, which result from a want of knowledge.

There are at least four different processes which are too valuable not to be described in this book.

The Morning Sponge Bath has already been described at page 13.

The Wet Sheet Pack, or *Full Pack*, is to be applied as follows :—Remove the feather-bed, if there is one, and then spread out two or three comforters so as to cover the whole bed, one over the other.

Then spread out over the comforters two or three
blankets. Then take a stout sheet, linen is best,
and wring it out of cold water so that it will not
drip, leaving it quite wet ; spread out the sheet
over the blanket, and then cause the person who
is to take the pack to lie down quite undressed exactly in the middle of the sheet, on his back, so
that the neck shall come just at the edge of the
sheet. The arms are to be placed against the
sides. Then throw one side of the sheet over the
person so as to envelop him all but the feet ;
spread it smoothly, and then throw the other side
of the sheet over the person, and spread that : then
commence with the blankets, and bring them
round the person *one side of one blanket at a time,*
tucking in each side alternately, *very carefully and
closely,* particularly about the feet and neck ; when
the blankets are all tucked in, bring over the comforters alternately, and tuck them in, in the same
way. A pillow may be placed under the head.
This should all be done *very rapidly,* after the patient first lies down on the wet sheet ; because the
shock and sensation of cold is pretty severe. This
however, very soon passes off, and a delightful
sensation of relief follows. This is especially the
case in fevers. If the feet are cold they should be
got warm before the pack is begun ; and a hot
brick be placed at the feet, partly within the pack.
If there is headache, the forehead, and the top of
the head, should be covered with a towel kept well
wet. The usual duration of the wet-sheet pack,
for colds, slight feverish attacks, and chronic complaints is one hour. The patient should then come
out and have a thorough sponge-bath, with much
rubbing with a coarse towel. The patient should
drink freely of water while in the pack ; or at

least enough to satisfy thirst. In case of violent fever the pack should be continued several hours until the fever is subdued ; it must be opened about every half hour, or hour, in such a case ;—or, as often as it becomes well warmed up ;—and then wet again by sopping in some cold water about the chest and body. The most violent attacks of fever rarely fail to be subdued in this way in six or eight hours ; and then the patient should come out, take a sponge bath, with much rubbing, and go into a dry bed. The proper medicine should be taken at the same time, as directed under *Inflammatory Fevers.* In some very violent fevers it is necessary to repeat this process, on some two or three successive days ; or as often as the fever returns. We have had experience of this in our own person.

The Wet-Sheet Half Pack is better for children, and those adults unused to water treatment. The comforters and blankets are to be arranged as for a full-pack : then a small sheet is to be wet and placed cross-ways of the bed, so as to reach from the arm-pits, nearly to the knees : the patient then lies down on his back, on the wet-sheet, and is to be enveloped as just described ; the sheet reaching from the throat, and arm-pits, nearly to the knees : and not forgetting the brick at the feet. The arms are left out of the pack ; which is an advantage for children, beginners, and sensitive persons; or, if it should be necessary to continue the pack many hours, as it admits of turning on the side. Never forget the caution to have the feet warm before going into the pack, which may be done in a few minutes by soaking them in hot water It is well to keep the hands warm also. In cold weather the pack should be given in a room

suitably warmed, but not over 65 degrees. *A
larger amount* of comforters and blankets is required to produce the necessary degree of warmth;
part of them can be unfolded, after the first feeling of chilliness has passed off. The pack should
never be allowed to get hot and dry; as it would
then do harm. If the above precautions are observed the Wet-Sheet Pack need no longr be a
bugbear; but will become, even with inexperi-
enced hands, a most powerful agent for good. For
young children, who have not been in the habit of
a daily cold bath, it is generally better to depend
upon frequent sponging the surface, in case of
fevers; and the medicines hereafter directed.

The Sitting, or Sitz Bath.

A tub of sufficient size, or a medium sized wash
tub will do. Place it about four inches from the
wall, and fill it so full of water that after the patient shall be seated, the tub will be nearly full,
and the water reach above the navel. The best
time for taking this bath is on going to bed: and the
proper duration from 15 to 30 minutes. If taken
on going to bed it is best to undress, but that is not
necessary.

One or two blankets, and a comfort, can be arranged so as to reach from the neck to the floor,
enveloping the tub and all; and a pillow may
then be placed between the back and the wall for
a support. The feet are to be left out; and if
they are inclined to be cold, should be placed on a
hot brick; and in case of headache, or heat about
the head, put a wet towel on the forehead, temples
and top of the head. It is generally well to rub
the abdomen briskly, during or after this bath;

which if taken on going to bed, will be found highly promotive of good sleep. It has the effect of
strengthening the nerves, of clearing the blood, and
humors, from the head, chest, and abdomen, and
of relieving pain and flatulency; and is of the
greatest value to those of a sedentary life. It may
be used by every person, whether in health, or otherwise, without the slightest fear of taking cold. It
is hereafter recommended for its most suitable complaints.

The proper temperature of the water is from 75
to 55 degrees. Let beginners commence at 75;
and go down one degree for every bath until they
arrive at 60 degrees. Persons much used to water may go as low as 55. Do not try to take baths
without a thermometer.

A hot sitz-bath, taken as hot as can be borne,
say at 110 degrees, will often relieve the most severe attack of menstrual or other colic. It may
be continued a half hour, or more; or until relief
is obtained.

A proper sitz-bath tub has a support for the
shoulders and back, and a place to rest the arms;
it may be made of wood or tin. Sitz-baths are
usually taken in the day time without undressing
entirely. A little experience will teach the most
convenient way.

The Sweating Process.

A mattress, or straw-bed, is to be covered with
two or three comforters, and then as many blankets; the patient should then be closely packed
in the blankets, and comforters, just as directed in
the Wet Sheet Pack, only there is no wet sheet.
If there is any difficulty about accumulating heat

enough to sweat, put two hot bricks wrapped up in cloths at the sides, within the pack, but not near enough to burn ; and one at the feet, which will rarely fail to accumulate heat enough to produce perspiration. It takes some time however ; and especially for the first time, the confined position, the irritation of the blankets, and the accumulating heat, will be found uncomfortable ; but not more so than other sweating processes.

The patient should drink freely of cold water, in frequent but small draughts, and more especially as soon as perspiration begins. It may require an hour or two for perspiration to begin, but the time may be shortened by the use of hot bricks, or bottles of water ; by *very close* and abundant packing : and by making motion with the limbs, as much as space will allow.

In most cases it will be better to pack in the wet sheet well wrung from warm water. Patients may take their choice ; and in all cases may take vapor, or steam baths, if they can get them instead. Any diseased parts, as swellings. or ulcers, should be bandaged with wet towels before packing up. Persons may sleep while sweating : but usually should keep awake to drink cold water. A window should be kept open during the sweating ; which, to do much good, must be profuse, and last as long as the patient can endure it ; or say from 45 minutes to an hour and a half. Patients may begin gradually.

On coming out *always* take a cold sponge bath, with a large basin full of water, and much rubbing. This will not only be agreeable but highly salutary.

Cooling Bandages.

These are small packs, on the principle of the wet-sheet. Make them with a wet towel, and a piece of flannel outside. Such a pack is very useful about the throat, in croup, and diphtheria. The inside bandage should be kept *wet and cool.*

Warming Bandages.

These are local packs, like the sweating process on a small scale. A towel is well wrung out, and kept covered with a dry cloth: as soon as dry it should be wet again. This is valuable in all dyspeptic troubles, and may be applied round the entire body, by taking a long towel, wetting one third of it, applying the wet end first over the abdomen, and then the dry part over, by passing it quite round the body. Fasten it as you like. If the kidneys, or back are in trouble, put the wet part there. It should soon become warm to do good. A woolen wrapper may be used if necessary. It will need renewing twice or three times a day.

Water Drinking.

Medicinally, this should be practiced an hour or two before a meal. In all dyspeptic cases from one to six tumblers of water may be drank before breakfast, or dinner with advantage. The idea that drinking much water *with* a meal is injurious, *is a mistake.* And in all cases where food disagrees with the stomach, drinking plentifully of water will assist in ridding it of its half digested load.

We may lay it down as a rule that always when

3

the body is not heated by exercise ; or is not much
fatigued, and there is thirst, we should drink until
satisfied. The very thirsty should drink slowly,
this will allay thirst safely and effectually.

NURSING.

Good nursing, we are sorry to say, is but little
understood ; and good nurses are scarce. The
position of nurse, is, in fact, a trying one ; and one
in which the failings of human nature are very apt
to appear. A good nurse should possess sufficient
physical powers, and therefore should not be old :
good judgment, and *some experience*, and therefore should not be very young ; take one, however,
that is too young, rather than too old. Good powers of observation, truthfulness, and honesty, are as
valuable here as in any other walk of life. Delicacy, and tact in managing the sick, are also important qualifications. Total abstinence from tobacco in any form, opium, and spirits, should be
expected of a nurse ; but quite frequently in vain.
Personal cleanliness, too, we shall sometimes look
for, and not find. Too much should not be required
of a nurse, lest she become exhausted, and fail
you in the time of greatest need. A good nurse
should not be headstrong, but of docile disposition,
that she may be willing to carry out directions,
even if contrary to her prejudices ; yet she should
have enough moral courage to insist upon obeying
the physician's directions, even if contrary to the
prejudices of the patient, or others. She should
be willing to follow directions implicitly, yet not
blindly : lest varying circumstances should require
slight changes. A nurse should not be officious,
or meddlesome ; for a good doctor should under-

stand, and order, everything about a sick room.
A sleepy and snoring nurse is a nuisance ; try and
make sure about *that* point before you engage one.

The question of what food and drink should be
given the patient, is never to be left to the decision
of the nurse and if the physician does not give
express directions they should be asked for. In
general, however, we will say that no good physician, at the present day, will prohibit cold water as
a beverage, in the sick room. It is always to be
allowed in small quantities, say a half a tumblerful,
at a time ; and often enough to satisfy thirst. It is
objected that vomiting, or purging is caused or increased by it: we say no matter; if caused by
water, that will soon cure itself. If the doctor re-
fuses a patient cold water, you would do well to
begin by dismissing *him*, and take your chance for
a better one.

Herb teas should be wholly banished from the
sick room as a pernicious nuisance ; and all warm
and hot drinks, except when specially ordered or
permitted, by the doctor, should be equally banished.

In regard to food it should never be crowded upon
a sick person against his wishes ; and in fevers it
is generally better to withhold all food while the
fever runs high, or until it begins to abate. The
very lightest kinds of nourishment only, such as
water of currant jelly, and toast-water, are allowable. All kinds of animal food, such as chicken
water or beef-tea, are injurious in high fevers ; and
should not be used, even for the convalescent, until it is ascertained that light vegetable food ; such
as toast-water, rice-water, and gruel, are well digested. Many a relapse has happened from eating too soon, and too much, during, and after a se-

vere illness. The doctor should specify the quantity, as well as quality of nourishment to be given. Among the *Receipts* will be found some well adapted to the sick; they are not generally of a nature to do any injury, if used moderately; but the doctor, if there is one, should decide. Ten sick persons are injured by eating, and drinking, where one suffers from going without. We have had the very best success with violent inflammations, and fevers, in many cases, where the patient has taken *only water*, for more than a week. Two weeks is not an excessive time to go wholly without food in violent diseases. The superfluous fat of the system is used up, at such times, as Liebig has shown, to support animal heat, and other vital processes. It is in this way that the sick lose so much flesh.

Cleanliness of the most scrupulous kind should be observed in the sick room, if in no other place. All vessels employed should be cleaned before they are used a second time; and those evolving odors immediately removed. The entire surface of the body should be sponged over with cool soft water, at least twice a day; and the face, chest, and body much oftener in fevers. The addition of any thing to the water used, unless as hereafter directed in this book, is quite useless. Such useless things are vinegar, rum, and spirits. The patient's mouth and teeth should be cleaned at least twice a day with a tooth brush, which he can generally best do himself. If not, the nurse should not neglect it. Once a day is better than nothing. A little fine white soap is as good as any thing in ordinary cases; but in fevers where the teeth are much incrusted, pulverized charcoal is better.

Quiet in the Sick Room need hardly be inculcated. Noise is frequently distressing to a sick person which he would not notice if well. The creaking of a shoe, or a door, or the rattling of a newspaper, may entirely prevent sleep, and aggravate fever. No whispered conversations, or whispering of any kind, are proper in a sick room. Say everything, in a quiet but cheerful tone of voice, which if awake, the patient may hear. If he hears whispering, he will think things are going wrong with him. No long conversations are proper; and no tales of any kind concerning unfortunate cases of sickness. The frequent opening and shutting of doors may be avoided, by keeping the door open as much as possible, which will favor ventilation ; or if it is the cold season, by having locks and hinges in good order, and oiled if necessary ; and to inform his friends when he is asleep, put the feathered end of a quill through the key hole, as they do in Paris. Visitors should be quite excluded from the very sick ; and no one should disturb the patient when asleep.

The ventilation of the sick room is of even greater importance than for persons in health. Enough has been said of the importance of *pure air* under the *Laws of Health* ; or may be found under Consumption. It should be secured by having a window, or windows, wide open, if the weather will possibly admit ; but at least a crack even in stormy weather. *Patients never take cold if their feet are kept warm ;* which is a point of great importance, especially in fevers. All evacuations should be immediately removed from the sick room, and *saved to show the Doctor.* Never burn anything *to cover up* bad smells ; it only makes the air still more impure. The only remedy is fresh air.

It is very injurious to keep the sick room too

hot; and always increases fever. A thermometer
is a necessary article, since without it, it is impossible to regulate temperature. This should not
exceed 65 degrees, but may be as low as 55, if
the patient's feet are kept warm. Nurses who are
constantly afraid of taking cold should be des-
pensed with if possible; they are rarely of the
right sort. A sick person is frequently overloaded
with covering, and thus kept feverish, when he
might perspire with a less amount. This plain
rule may be given;—if a sick person complains
of too much heat, attempt to remedy it, either by
tepid sponging, or by removing a part of the cover,
or both.

Sick persons are often made worse by improper
attempts to sit up. Sitting up, after illness,
amounts to exercise. Its amount should be in
proportion to the strength. The great error is
always on the side of too much, from a mistaken
idea that sitting up is the only way to gain strength.
After a long, and severe illness, fainting may
easily result from sitting up even a few minutes.
The sick person should generally be the judge,
and decide by his feelings as to whether it is agree-
ing with him or not. Be careful not to overload
a person with covering when sitting up. It is
equally necessary to keep the patient out of a
draft of air : but do not shut all your windows,
and doors, because he is sitting up.

The making of the Bed, in a sick room, is generally badly managed. A very sick person should
never sit up to have the bed made ; but be placed
in another bed, or on a sofa, while it is being done.
Frequently when mattresses are used, one side
may be made up with fresh linen, and the sick one
be removed to it, while the other half of the bed

is then renewed. This is the best way when the
bed is wide.

As a rule, the bed and body linen of a very sick
person should be changed *every day*; and this
should be done in the morning. It conduces highly
to the comfort of the patient, and the success of
the remedies. A mattress in case of fevers, and
all acute diseases, is greatly to be preferred to a
feather-bed. It is an advantage to a very sick
person to change from one bed to another, every
12, or 24 hours, whenever it can be managed.
In that way thorough renewal, and airing of
the bed is secured. When on a wide bed, the patient should be changed from one side to the other,
as often as he wishes; or several times a day, especially in warm weather. Whenever any change
is made do not fail to see if the feet are warm:
for that is a great sign of safety.

A bed-pan and urinal are quite essential for
managing the evacuations of very sick persons;
since the strength is very much saved in that way.
Such patients should by no means be allowed to
get out of bed to perform their evacuations, when
their strength can be so easily husbanded. A good
nurse will manage all that very skilfully ; a bad
one will bungle it.

Relapses during convalescence take place more
frequently from over-eating than from any other
cause. Many a convalescent person has eaten himself into a dropsy. The change from the sick
man's best diet of pure water, to his usual fare,
should be *very gradual.* Animal food of every
sort, including broths, and jellies, should be eaten
very sparingly, if at all, until the strength has
very much returned. Wheat, or oat meal jellies,
puddings, or porridge are among the lightest, and

most nutritious kinds of food. Persevere in their
use and there will generally be no risk of relapses.
The entire wheat is peculiarly suitable for conval-
escents, either in the form of bread made of wheat
meal; or boiled, like rice; or in puddings. Milk
may be usually allowed with it, after convalescence
is fully established.

SCROFULA.

Scrofula, or *King's Evu* begins in the blood, and
as this vital fluid penetrates every pore of the body,
it carries with it the contamination of this corrupting disease. Scrofula very much resembles the
poison, virus, or infectious taint commonly known
as Syphilis, or *the pox*; the resemblance in many
cases is so striking that good physicians are puz-
zled to tell with certainty to which disease of the
two, these cases, or symptoms belong. The reason for which is that in very many cases the two
are mingled, or combined together in the same
system, being derived from a parent, or grand-pa-
rent, by what is called hereditary transmission.

The taint of Scrofula, acting like a hidden poison,
after corrupting the blood, and being engendered
within it, so spoils, and as it were rots out the
glandular machinery of the body, that those vital
organs can no longer do their accustomed duty.
Hence the liver, the kidneys, the lungs, the spleen,
and the lympathic, and other glands, are no longer
able to clear the system of its waste and worn out
materials. Hence the blood remains burdened
with corruptions, by which the entire strength, and
vital energies are weakened; and thus persons
afflicted with this taint, become much more subject
to other diseases.

In many cases it may long lurk without any very decided, manifest, or ulcerous signs; but in such cases is more likely to produce Pulmonary Consumption; which is only Scrofula located in the lungs. Nearly one third of all the deaths take place from Scrofulous diseases including consumption.

It is not necessary here to enumerate the causes of Scrofula, as they may be found very fully detailed with those of consumption, a little farther on in the book. It is only necessary to say that no able medical man doubts the fact that this disease descends from parents to children ; and the opinion is fast gaining ground that the contamination of Syphilis is frequently combined with it.

The Signs of Scrofula.

The chief signs, symptoms and indications of Scrofula are ;—a pale bloated face, indicating a lack of energy, often fair, with a transparent whiteness of skin ; and agreeable redness of cheeks; sometimes a sickly yellow hue round the mouth ; livid circles round the eyes; pearly whiteness of the whites of the eyes threatening consumption ; whites of the eyes red and injected, threatening dyspepsia ; the eyelids are often humory and run ; or are red and inflamed ; the teeth are very white, liable to crack, and decay early, or are foul and covered with a greenish or glairy slime ; the appetite is unnatural, sometimes bad, and often too good ; the tongue is foul, and the breath bad, nausea and sick-headache, sour stomach, piles, and constipation, or diarrhœa, are common ; the flesh and muscles are flabby ; the limbs are soft and full, but lack firmness ; often considerable loss of

flesh and emaciation; general feebleness, and debility; the powers and functions of body and mind
though feeble, are often too early developed; a
scrofulous child grows rapidly; the bones are often
weak, crooked and small, producing deformity, or
lack of symmetry; *eruptions and diseases of the
skin are very common*: the glands of the neck,
arm-pits, and other parts swell, and then inflame
and ulcerate: *humors and eruptions of the scalp are
very common*; there is always a want of robust and
vigorous health. The principal affections of the
skin which accompany Scrofula are Erysipelas,
Pimples, Blotches, Boils, Tetter, Salt Rheum, Tumors, Scald Head, Ringworm, Ulcers, and Sores.

Not all of the above signs, symptoms, and disorders will be found in any one person; but *any* of
the above-named peculiarities indicate the hidden
and lurking contamination, which at any time
may break out in open eruptions and sores, or attack the lungs in the form of consumption. All
other diseases occurring in such persons become
more obstinate, and the process of cure slow.

We are indebted to the noble science of medicine for the discovery of a remedy capable of eradicating this foul disorder from the blood, and entire system. It far exceeds any other known medicine, not only in its power over scrofulous disease,
but in the quality it possesses of purifying the blood
from every sort of taint, and restoring it to a
healthy condition For a further account of the
discovery, and perfection of this remedy, see the
next article but one. It is called Rush's Sarsaparilla and Iron. This medicine should be taken for
any and all of the conditions and symptoms just
described, in the dose of a teaspoonful, immediately after breakfast and tea, and on going to bed,

and two teaspoonfuls after dinner. After taking it a week, increase the dose a little by taking two teaspoonfuls after each meal, and one at bed-time. Do not take it in larger doses than directed; for though it would do no serious injury; it would be wasting the medicine.

Scrofulous persons should read over, and follow as strictly as possible, the directions, and hints, given under *Laws of Health, and Prevention of Consumption.* The same causes which produce consumption, produce scrofula, and the same rules in regard to *pure air* and exercise, apply to both. While taking the Sarsaparilla and Iron, scrofulous tumors, sores and ulcers should be kept dressed with pieces of old linen, of the size of half a towel, or less, wet in pure soft water, in which has been put a little tea made of walnut leaves by boiling, say a teacupful of leaves in a pint of water: keep this infusion in a cool place, and use a teacupful of it in a quart of water to keep the linen cloths wet with when the sores are dressed. These should be bathed twice a day with the same wash. The wet cloths may be covered with a flannel bandage, or covering. If any greasy application is made let it be only a piece of linen thinly spread with mutton suet freshly melted. The only good which that will do, is to keep out the air.

Most cases of sores and diseases of the skin yield very promptly to the use of Rush's Sarsaparilla and Iron; yet some cases may be so deeply seated in the blood, and glandular parts of the system, as to require great perseverance in the use of this most valuable remedy. In general the longer scrofulous signs, and troubles have been manifest, the longer must the medicine be perseveringly taken.   Three to six months, or even a year, will sometimes be required before perseverance will be rewarded by complete restoration to health. Scrofulous sufferers are apt to get out of patience, and change about from one remedy to another, until the stomach, and whole system is much injured by pernicious drugs.   A better way is to consult some able physician in whom you can trust, and who has had *success* in treating such complaints.   The editor is in constant receipt of letters asking advice in obstinate cases, and at the request of influential friends has concluded to give his address in this book.

THE EDITOR'S ADDRESS.

Persons wishing for information, or advice, on any medical subject, but more especially on diseases of a chronic, or virulent nature, may address, A. H. Flanders, M. D. Lowell, Mass.   All communications are promptly answered, and advice is cheerfully given free of charge. Medicines are often sent by mail, or express: and in order to facilitate a full understanding of the sufferer's case, a printed list of questions is sent to him, or her, to be filled out, when it is supposed treatment may be desired.   Please send a stamp to pay return postage, and write your address, post office, county and state, very plainly.

It is always preferable to see the editor in person, at least once, especially in cases of consumption, and female complaints.   He may always be consulted *free of charge*, at his rooms in Wentworth's Building, Lowell, Mass.

THE PHYSICIAN.

Thousands and tens of thousands of most precious human lives are annually sacrificed to the iron despotism of theory ; to an obstinate and blind attachment to system ; to a stupid, insensate, and mechanical routinity, which treats all cases alike, because the symptoms are similar—when, in fact, there are no two cases precisely alike, even where the cause or the primary disease is the same. As all human faces differ, so do all maladies of the same name, nature, and location, differ in different persons. Many diseases, having almost the same symptoms, differ altogether in character, and arise from entirely distinct causes ; hence, each case should be examined, studied, and treated for what it is really in itself—as no one individual, or any number of individuals, should be made the absolute standard of any other case, nor the treatment be made the exact standard of any other.

A physician, to be successful, must be, so to speak, a UNIVERSAL man—one capable of appreciating the habits, constitution, and condition of the people of all classes, of all countries, the temperaments of the young and old, of men and women—able to accommodate himself to the states of all, as readily recognizing the signs of health or disease in the ignorant as in the learned, in the fastidious as in the frank, in infants and children as in the adult. His knowledge should be universal. He should know everything which can be known of the human economy, and the differences between all varieties, ages, sexes, &c. ; and also everything which affects the human system, either favorably or unfavorably—all of air, moisture, climate, temperature, dress, food, medicine, good and bad—the effects of all the various occupations, the

gases, acids, woods, paints, oils, minerals, and earths
which are used in mechanical operations and the
arts—the effects of the positions, places, and all
other circumstances in which men, women, and
children labor, live, or are educated—the effects of
all the various modes of training, the studies, cares,
anxieties, troubles, passions, emotions, professional
pursuits. recreations, associations, and every other
particular relating to body or mind—that thus he
may be able to detect the hidden cause of the disorder which he may be called on to treat.

Large experience, a profound acquaintance
with the human constitution, iu health and in disease ; a genius, and instinctive aptitude for the pro-
fession : a ready, and almost intuitive perception of
all the variation from a healthy condition ; an intense interest in, and cordial sympathy with every
case of suffering ; a conscientious regard for the
well-being of all, regardless of wealth or poverty,
of intelligence or ignorance, of high or low condition, which shall ever induce us to do unto others
as we would they should do to us ; and above all,
an undoubted reliance upon a higher Power, who
is able to make us apt and wise in devising ways
and means, when ordinary resources fail us—all
these are indispensable in the man to whom the
lives of multitudes of human beings, of every age,
and of both sexes, are entrusted. To penetrate
deeply into a case, so as to see it truly, and treat
it successfully, we must enter into the interests,
and feelings of the patient — we must make
his case our own—bringing all our knowledge, experience, ingenuity, untiring energies, and best
feelings, to bear upon it—at the same time cher-
ishing a firm confidence of ultimate success, where
there is any ground whatever to hope for a cure.

RUSH'S SARSAPARILLA AND IRON.

This is a highly concentrated preparation of the best *Sarsaparilla* imported ; *chemically combined* with the most delicate, and easily vitalized form of *Iron* known to modern science.

The properties of the Sarsaparilla of Central America to remove, and eradicate from the system, all the various forms of scrofula, have long been known. Its mode of operation is by purifying the blood, and thereby restoring its healthy condition, and properties. This would seem a very simple process, and Sarsaparilla is no doubt alone suffic- ient for this purpose, where the supply of *Iron* in the blood is already sufficient. But unfortunately for the entire success of Sarsaparilla *alone*, the blood in most Chronic Diseases, and in all diseases of Debility is considerably deficient in this Iron, which is its most necessary ingredient, when in a healthy state. Liebig, the greatest analytical chemist of modern times considers that the iron in the blood is the main agent in carrying on that very vital operation of renewing it, in the lungs, by its conversion from dark-colored, or venous blood, into bright red, or arterial blood. He says that this proportion of Iron is in the form of a *protoxid* : and is converted in the lungs, into a *peroxid*, by taking up oxygen from the air which we draw in at every breath. However this may be, it is cer- tain that in all diseases characterized by *debility*, loss of flesh, unnatural paleness of face, wasting away of the muscles, and every form of Scrofula, Consumption, Secondary Syphilis, Dyspepsia, and Anæmia, or Consumption of the Blood :—this vital fluid is deficient in Iron. The researches of those eminent physiologists, Carpenter, Longet, and others confirm this. *No respectable physician will deny it.* Why not, then, the reader asks, give Iron in all these cases ? We answer that most physicians, though not disposed to deny the theoretical value of Iron, have been unable to find any preparation sufficiently delicate to agree with sensitive persons; and to be taken up, and readily assimilated, or united with the partially diseased, and disorganized blood. Hence when such crude prepartions as the *Carbonate of Iron*, *the Muriated Tincture*, *Sulphate of Iron*, and the usual drug-store preparations, are taken ; *headache*, and other disturbances occur ; because there is not power enough in the stomach, and other vital organs, to cause these crude drugs to assimilate, and combine with the mass of the circulating fluid.

There is another reason, also, why iron *alone* in *any* form, has not the power to cure the diseases before mentioned. It is because it has not the power to *purify* the blood, by eradicating the Scrofulous humor, and other morbid humors, which are there first engendered, and produced. The Sarsarilla *has* this property. No one who has ever had any large experience in its use, in such diseases as before mentioned, can doubt its purifying, and healing nature. The celebrated Professor Rush speaks in his lectures, and published works, in the highest terms of this medicine ; and has left a form for preparing a mixture of it, with Iron, which he used to order in almost all chronic diseases. There is no doubt his great reputation, and success, in the treatment of such disorders *was mainly owing to* the use of these two remedies together ; but he never succeeded by his formula, in making the two *combine chemically*, or form anything more than a *mixture*. . Chemical science, at that time, was not suf-

ficiently advanced; and it was left for the editor
of this book, while professor of Chemistry at Philadelphia, to make this discovery, by which the *active principle* of the Sarsaparilla is made to *combine
chemically* with Iron, in its most highly magnetic,
and minutely divided state.   The benefit of this is
that the nauseous and bitter taste of the Sarsaparilla, when *concentrated*, is wholly avoided; and
this *chemical compound* becomes so pleasant to the
taste, that the most refractory child takes it without a wry face, and it agrees with the most delicate,
and sensitive stomach; never causing headache,
and other disturbances, which *all* other preparations of Iron, and even the original formula of Dr.
Rush, are liable to do.

Chemical Analysis of healthy blood shows the following constituents in 1000 parts.   Water 790;—
Red disks or globules (coloring matter containing
the iron) 127;—Albumen (like white of egg) 80;
—Fibrin 3;—total 1000; but in all diseases of
debility, the watery part increases, and the red disks
containing the iron diminish.   According to eminent Physiologists, *these red-disks have the power of
reproducing themselves,* and are therefore the *very
life* of the blood; but this wonderful process is impossible without a suitable supply of iron.

It will be seen from the above analysis, that the
globules or red-disks form but a *small part* of the
entire mass of the blood, which in a healthy man
weighs about thirty-five pounds, or four gallons.
Of the blood disks only about 1-100th part consists
of iron.   But we must remember that every pore
and minute vessel, not only of the lungs, liver,
and brain; but also of the skin, and every other
part, is permeated, and invigorated by these minute
blood-disks, whose average diameter is only 1-

4

3400th part of an inch. We have treated of the importance of taking the iron as a medicine, in a delicate and easily vitalized form ; and the comparative *uselessness* of taking it, in a crude and bulky state.

The principal disorders, and states of the system for which this chemical preparation is curative, and valuable are as follows;—*Scrofula*, in all its forms; *Diseases and Eruptions of the Skin ; Anæmia or thinness and poverty of the Blood , Syphilis , Fever and Ague , Debility. and weakness*, from long sickness, wounds, or loss of blood ; many cases of *Consumption. Dyspepsia, Female Weakness, and Leucorrhœa or Whites* ; *Neuralgia, Pain in the Bones, Rheumatism ; Boils and Tumors ; Flatulence and Dropsy.*

The principle signs and affection of Scrofula, for which this medicine is specially adapted, are ;— *Humors, Eruptions of the Skin, and Scalp*, Salt Rheum, Tumors, Tetter, Erysipelas, Ring-Worm, Swelled and sore Glands, Pimples, Blotches, Bloating, Sore-Eyes, Inflamed Eyelids, Broken, or Early Decayed, and foul Teeth, Deformed, and Misshaped, Body or limbs, Pearly Whiteness of the Eyes, Pale Bloated Face, or very Fair transparently white Skin, Flabby Flesh, and Muscles ; circle round the Eyes. See page 40 of the Family Physician. The principle signs of *Dyspepsia* are ;—Sick headache, dizziness and confusion of head ; face pale, or yellowish ; bad taste in the mouth ; want of appetite ; sickness at the stomach; bitter belchings and vomitings; heartburn ; sour stomach : flatulence ; fullness at the pit of the stomach ; tight feeling of the clothes at the waist ; cramps and pains in the stomach ; red urine, with brick-dust sediment ; sleep restless and unrefreshing ; Constipation and Piles.

It is not thought best to encumber this book with a multiplicity of certificates, and letters, showing the nature of cures performed. We shall, however take room for the following letter.

A Cure of Scrofula.

201 Camden St., Philadelphia July 2, 1864.
Professor Flanders, Respected friend :—

You request me to state what is now the condition of my health, and if I am so disposed, what it was when you were here. I will do so very cheerfully : and hope that others will profit by my example. At the time you was Professor in the Medical College here, which I think was about 1856, I had been very bad with the Scrofula for nearly four years, had the humor from a child, inherited from my mother, and had spent a great deal of money for medicine and doctors, but grew worse rather than better. I had swelled and ulcerated glands in my neck, and on my limbs, sometimes an itching and burning tetter, on my hands and arms, and sometimes on my face, and in my hair. I could cure it, or dry it up, in one place, and then it would break out in another. The doctors said if I *did* cure it on the outside, it would only strike in, and turn into Consumption ; so I dispaired of being better, till one day I was at the College and you told me to take old Dr. Rush's Sarsaparilla and Iron. I had always used his Pills, and liked them, and so determined to do as you said. In two months I took two bottles, and was very nearly cured. The third quite cured me, and I was so when you went to New York. Three years ago I had the glands again swell in my neck, and took one more bottle, and have been quite well ever since. You ask if I know of any others that have been cured as I have. I know of three others that have been cured by the Sarsaparilla and Iron. They were sick some as I was; one of them was worse. I know of two that has been cured of Consumption by the Lung Balm. They were very bad and the doctors said it was consumption. I can get their names and residences for you, and should like to, you was so good to me. I shall never forget your kindness.

JULIA FENTON.

Other cures might be cited, and we give the following.

Henry W. Brown, of this city, applied to the editor about seven months since. At that time he had a running sore on his leg which had been discharging a great deal of scrofulous matter, for nearly a year; he also had sores in his neck, which broke out one after another; and was subject to humors, and pimples, and blotches all over his body. We recommended him to take Rush's Sarsaparilla and Iron, which entirely cured him in about three months. The improvement began very soon after he commenced taking it. When he began to take it he weighed only 110, and now weighs over 160 pounds. It should also be stated that he had a bad cough and reasonably feared that it was settling on his lungs. It was cured by a bottle of Rush's Lung Balm.

Before he came to the editor he said he had taken no less than nine bottles of different kinds of Sarsaparilla, and other medicines, which had done him no good.

Further certificates and letters relating cures may

be seen by calling on the editor at his rooms in Wentworth's Building. Let no invalid suffer for want of light and knowledge.

Rush's Sarsaparilla and Iron is for sale by most druggists, but when it cannot be procured in your neighborhood you can always obtain it of the editor by writing to him, and enclosing $1 for each bottle required; when it will be sent paid, by express. The editor's address is given at page 44. Each bottle contains 130 doses, or enough for cne month; and this is not too much to secure any good result. It is therefore the cheapest Sarsaparilla in the market; as other kinds only contain about 50 doses. It should be taken according to the following

Directions for Rush's Sarsaparilla and Iron.

Take a teaspoonful after breakfast, and tea, and two teaspoonfuls after dinner. After taking the medicine one week, take two teaspoonfuls after each meal.

Persons who are restless at night, will often be relieved, and sleep well, by taking a teaspoonful in a glass of water, on going to bed.

Children under five years of age, *six drops for each year of age*; children from 5 to 10 years old, 40 drops; 10 to 15 years old, a teaspoonful, and over 15, the full dose: it is pleasant to take; and may be taken alone, or in a little water. You will find in the Family Physician directions for using this medicine for each disease. Look in the Index at the end of the book.

RUSH'S LUNG BALM.

The original preparation of the great Dr. Rush, which he familiarly called his Lung Balm, or Balsam, was the best remedy for coughs, colds, and all disorders of the throat and lungs; and it has never yet found its equal as a general medicine for all that class of disorders. It was the principal means of success upon which that eminent man relied in such cases, and added greatly to his reputation. But it was a great source of regret to him, that the Chemistry, and Pathology, of his day was not sufficiently advanced to make the discovery of any very reliable remedy for *Consumption* possible. The Lung Balm as he then prepared it, was a certain remedy for the cough, expectoration, night-sweats and other ordinary *symptoms* of that disease. but had no power to remove *tubercle*. Indeed the entire Chemistry and Pathology of Tubercle was then but little understood. The great Dr. Rush, however, had the right idea, and suggested in one of his most learned works that when a curative remedy for consumption should finally be discovered, it would be found to have the property of *dissolving tubercle*. Bearing this suggestion in mind, we further entertained the idea, that it must be from the want of some necessary constituent of the blood, that *Tubercle* arises. This conclusion seemed borne out by the acknowledged fact that *most diseases of debility* are from a want of a due quantity of Iron in the blood; without which healthy blood can not exist, and which being supplied in such diseases, a cure follows.

On full consideration of all the elements which Chemistry shows to enter into the composition of the human body, *Phosphorus* seemed as most probably lacking in the system and blood of *tuberculous, or consumptive persons.* This seemed the more probable from the fact that this substance plays a highly important part in the animal economy, entering not only into the constitution of the blood, and the brain; but forming as the Phosphate of Lime, the greater part of all the bones of the body. *Chemical analysis of tubercle* also convinced us that its earthy part, or that which gives it a gritty feeling and substance, was only an impure concretion of Lime; the Lime belonging in the *bones,* and not in the *lungs.* The problem therefore we conceived to be, to administer Phosphorus as a medicine, in such a form as to be *harmless,* and at the same time have it combine or unite with the *Lime* of the tubercles, and carry it out of the system, or *transfer it to the bones where it belongs.* THE TUBERCLES WOULD OF COURSE BE DISSOLVED, AND DESTROYED BY THIS PROCESS, and thus the disease cured.

Much space and time would be required to mention all the different combinations of phosphorus which we prepared and used, in pursuance of this idea, all of which we found beneficial in consumption, particularly the Hypophosphites of Lime and Soda. But it was only after trials of more than four years (1855 to 1859), that we were able to decide that the Hypophosphite of Iron far exceeded any other chemical compound *in its power of dissolving tubercle.*

It was also found that the most effectual mode of administering it to consumptives, was in combination with the Lung Balm of Dr. Rush: and experience in its use soon established the exact proportions of each to be used, in Rush's Lung Balm as now made   ,

We had no disposition to conceal this light under a bushel, but encountering the indifference, or
ridicule of the faculty, as did the discoveries of
Harvey and Jenner; and wishing to test it thoroughly for several years, we have not decided, until recently, to offer it to all consumptive sufferers
everywhere.

Rush's Lung Balm, therefore, as now prepared,
possesses this power of dissolving tubercle, and at
the same time affords the most reliable and pleasant relief to the permanent symptoms of Consumption, of which it proves curative in the first
and second stages, in the great majority of cases,
wherever large cavities have not already formed,
and enough of the lungs are left to carry on the
vital process of Respiration. And even when
large cavities have formed, experience shows that
cures are very possible. We are satisfied that
there is comparatively little hope for consumptives under any other plan of medical treatment.

Under the combined action of RUSH'S LUNG
BALM and Rush's Sarsaparilla and Iron, the consumptive soon realizes a constantly progressive improvement in his system, with a rapid change and
amelioration of the more painful symptoms of the
disease. The spirits become more buoyant; the
the appetite improves: the patient increases in
flesh and strength, and the soft bloom of returning
health takes the place of the hectic flush. The
nervous system is also invigorated: the eye no
longer gleams with a false lustre; and the muscular movements instead of being feeble and languid,
become animated and confident.

The use of Rush's Sarsaparilla and Iron, at the
same time with the Lung Balm, is advisable in all
cases where there is little or no feverish action.

It assists very much in eradicating the Scrofulous, and tuberculous taint from the system. See page 47 for the virtues of this remedy, It should be taken as there directed. Consumptives should not fail to read the article on consumption, and profit by the suggestions there made, where directions will also be found for the use of Rush's Remedies. Rush's Lung Balm is for sale by most druggists, but when it cannot be procured in the patient's neighborhood it can always be obtained of the editor by writing to him, and enclosing $1.30 for each bottle required; when the medicine will be sent by express *free of charge*. The editor's address is given at page 44. Each bottle of Rush's Lung Balm contains 130 doses; or enough for one month. It is therefore the cheapest medicine in the market. It should be taken according to the following.

Directions for Rush's Lung Balm.

Consumptives should take 35 drops every six hours; and that is the dose for coughs and colds, for all adult patients, or those over seventeen; persons under 17 years of age may take *two drops for each year of age.* This medicine is pleasant to take, and may be taken alone, or in a little water. Shake the bottle a little before using. Directions may be found for each separate disease by looking in the index of the Family Physician at the end of the book.

CURES OF CONSUMPTION BY RUSH'S LUNG BALM.

Case of Harriet H. Palmer, Brooklyn N. Y.

[Copy of Certificate.]

No. 87 Grand St., Brooklyn, July 16, 1861.
This is to certify that I was sick in the year 1859 and 1860, and was given up to die by two doctors. They sounded my lungs and said I had tubercles. I have lost two sisters by Consumption and thought I was going to die. I work on coats and vests, but had not strength enough to earn much, when I began to take Rush's Lung Balm, and other medicines recommended by Professor Flanders. I had night sweats very bad, and coughed up a great deal of bad looking matter. I began taking Rush's Sarsaparilla and Iron, and the Lung Balm in the Winter and used the Inhaler too. I also used a little of Rush's Restorer in water for bathing, and it kept me from taking cold. I took about 8 bottles of Rush's medicine, and got quite well in the summer of 1860. The two first bottles of Lung Balm helped me a great deal. I verily believe that I owe my life to these medicines. I am now as well as ever I was in my life.

HARRIET H. PALMER.

She also had some spitting of blood, but has omitted to mention it in the above certificate.

The following case was reported to the editor, and first published, in the Journal of Medical Reform, for January 1861.

Case of Lydia Grimes. Williamsburg, Long Island.

Miss Grimes, born of a consumptive mother in 1839, had always been delicate. In Dec. 1859 she had an attack of bleeding at the lungs, and previously a bad cough for which, she said she had taken all the advertised cough, pectoral, and pulmonic syrups without relief. She was then attended by Dr. Wright through January, and February, who very much relieved her cough, and other symptoms, till March. At that time while engaged in putting up some curtains, she was seized with another attack of bleeding, and after that grew rapidly worse. Her cough returned, and with it much expectoration of greenish and yellowish mucus and matter, often streaked with blood. She also had *Hectic Fever* as indicated by fever turns, and profuse night sweats. She had lost flesh very much, and weighed but 91 pounds. At this time she was unhesitatingly pronounced to be in a pulmonary Consumption by Drs. Wright and Arnold; and becoming discouraged with the regular faculty, she went to a Philadelphia physician, who came to see patients at one of the New York Hotels, and made examinations with the respirometer, &c. She was intending to be treated by him, but not having money to comply with his exorbitant terms, concluded soon after to take Rush's Lung Balm, which was recommended to her by a friend who had been cured by it. She was so much benefited by the first bottle that, she was induced to persevere, and was about this time first visited by the editor of this book, who further prescribed Rush's Sarsaparilla and Iron, and the use of Rush's Restorer in bathing the chest and body, to assist in preventing the night sweats. The remedies were continued in this way through April and May with great benefit. The night sweats during this time had nearly ceased, and she had begun to improve in flesh and strength. The cough too was very much better. In June she left off the Sarsaparilla and Iron, but continued the Lung Balm, the bathing, and the Restorer. During the Summer and Fall the improvement went on, without any drawback, except one attack of raising blood, and by November she was as well, so she said, as she ever was in her life. She then weighed 127 pounds. This cure has continued permanent to all appearance; but some weakness of the lungs has always remained, and probably always will, which brings on a cough if she takes cold; but this is always promptly removed by a few doses of Rush's Lung Balm, which she kept on hand for the purpose.

I have read over the above statement of my case, and can say it is all true; though I think I was sicker than the Doctor says.

LYDIA I. GRIMES.

Kings County, ss. Subscribed, and Sworn to, before me, this 23d of November, 1863.

WM. P. STONE,
Notary Public.

Case of Thomas Clarke,

LOWELL, MASS.

He is a leather-dresser by trade, and relates that he lost both father and mother, and a brother by consumption. Last November (1863), he was

attacked with Lung Fever, for which he took the
usual remedies, but suffered a relapse which left
him with an obstinate cough, with which he raised a
great deal of matter, which was often streaked or
spotted with blood. This continued in Dec. and
Jan., during which time he became very much reduced in flesh, and had frequent night sweats. He
relates that he took various pulmonic and pectoral
remedies and balsams, but grew worse in spite of
them, as well as the usual old fashioned medicines,
which he took after he had the lung fever. In this
condition, receiving no encouragement from physicians who examined his lungs; and the cough, expectoration, and night sweats being very bad, he applied to the editor. We found from a thorough
examination of the lungs, with the stethoscope, and
by the ear, that he had tubercles, and a cavity
at the top of the left lung. He then commenced
taking Rush's Sarsaparilla and Iron three times a
day, and also Rush's Lung Balm every six hours
for the cough; he also used Rush's Restorer in
bathing the chest, which very much helped in
checking, and curing the night-sweats. He very
soon began to improve in appetite, which had
nearly failed him; and his cough was at once relieved of its most distressing symptoms. He suffered a slight relapse in March, from exposure, but
after that continued to improve regularly. He
took but one bottle of Sarsaparilla, but continued
the other medicines through the spring and summer. He said the Lung Balm did him the most
good of anything; but it was apparent to us that he
also derived benefit from Inhalations of Rush's Restorer. He continued to improve, and gain flesh,
until about August 15th, when he called himself
well, and said he needed no more medicine. This

entire cure occupied, therefore, a little over six months.

We have on hand a large number of letters and notes, showing cures of Consumption, Asthma, Heart Diseases, &c. &c.; either completed, or in progress, which are freely shown to invalids; let any, who desire, come and be satisfied that all diseases are curable, when rightly and seasonably treated. None should perish for want of light and knowledge.

For testimony in regard to two cases cured by the Lung Balm, see page 52.

Opinions of the Press.

From the Journal of Medical Reform, July, 1, 1864.

" We have already had occasion to speak of the cures of scrofulous, and consumptive diseases, effected by Rush's medicines, and more especially the Lung Balm; one, or more cases have already been reported in this Journal, and two more have recently come to our knowledge. It would almost seem as if a remedy had at last been discovered for consumption; and that the cause of medical reform is to be materially aided by the popular use of Rush's medicines."

From Hull's Herald of Health for May, 1864.

" Through the politeness of Professor Flanders, the proprietor of Rush's medicines, we have just received a package of Lung Balm, and Sarsaparilla and Iron. We are satisfied that they are valuable remedies, for the disorders for which they are recommended. It is an abuse of terms to call such medicines as these, 'quack' medicines."

RUSH'S CATHARTIC PILLS.

This is the safest, mildest, and at the same time most effectual family physic pill now in use. It is a powerful promoter of the action of the Liver; and entirely avoids the necessity of calomel, or blue-pill. It is entirely vegetable; and being prepared from the juices of plants, and roots, which grow in our own forests, is especially adapted to the system of our people. This pill is not thickly sugar coated; but *as much as is consistent with the preservation of the medicine.* A *thick* coating of sugar, in many cases, *injures the medicinal effect.* This pill is also the most economical in the market; *three* being sufficient for a dose; while *seven* are frequently required of the sugar coated pills.

Families who once try these pills will not exchange them for any other. They are in short the *very best* pills for every purpose for which a cathartic, stomach, or liver pill can possibly be used.

DIRECTIONS. Take two pills, and in case there is no operation in four hours, take two more. Difficult persons may take three or four. Commonly take a dose at bed-time, and again early in the morning, if necessary. Price 25 cents a box; sold by all druggists.

Caution.

The great popularity of Rush's medicines has induced some unprincipled persons to make, and sell, entirely worthless imitations. To guard against such imposition, observe that the name of the proprietor, A. H. Flanders, M. D. is stamped in the glass, of every bottle of Sarsaparilla and Iron, and Lung Balm, and his initials on every bottle of the Monthly Remedy. Every box of Pills, and Restorer, has his signature on the box. None others are genuine.

PULMONARY CONSUMPTION.

Consumption of the Lungs, Phthisis.

No fell destroyer ever casts athwart the family threshold and the sacred affections of the home circle a more lurid, hopeless, and ghastly shadow, than does that cankering, insidious, and remorseless foe, CONSUMPTION. It is the DESTROYING ANGEL of our climate, and, of all diseases, is the most frequent and the most fatal. It claims as its victims nearly a third of all who die, and, regardless of age, sex, or condition, it wastes equally by noonday and by darkness; and preys alike upon the tenderness of infancy and childhood; upon the bloom and strength of the young and gentle, the loved and lovely, the mighty and mature, and the helplessness of declining years; and bidding defiance to the powers of science and the skill of medical men, and knocking impartially at the cottar's humble abode and the royal palace, sweeps its countless millions into " the dark valley of the shadow of Death." How strangely illusive the brilliant flush mantling the cheek with the earliest blossomings of the bloom of decay ! What a bitter mockery to hope and sense is the bright, electric, unearthly sparkle that illumines the eye with a preternatural light, too intense for health or length of life.

Such has *heretofore* been the almost unvarying course of this disease ; but in the hope of being

64

able in a later portion of this work, with the help of modern chemical and other science, to make known certain remedies by means of which this disorder has been, in many cases, shorn of its power, we shall now proceed to give a detailed account of it.

The word CONSUMPTION — from *consumere*, to waste away — in its most extended signification, as applied to the living body, denotes that progressive emaciation which usually precedes death in most chronic diseases. The appellation is, however, more commonly used to designate a *scrofulous ulceration* of the lungs. By consumption, as used in this country at the present day, is generally understood a WASTING *away of the substance of the lungs.*

Dr. Rush, in his medical dictionary, makes seven varieties of consumption. 1. Incipient consumption, without an expectoration of pus or matter. 2. Consumption with an expectoration of pus. 3. Consumption from scrofulous tubercles of the lungs. 4. Consumption from bleeding from the lungs. 5. Consumption from eruptive diseases. 6. Consumption from chlorosis of females, or disordered uterine action. 7. Consumption from self-abuse ; or from venereal ulceration of the lungs. But we must remember that *pulmonary consumption* is only a part of a great constitutional malady, which attacks most conspicuously the lungs. This constitutional disease is the *scrofulous,* or, as it occurs in consumption, the *tuberculous* taint.

Scrofula, as we have already remarked, is an hereditary evil ; so also are *tubercle* and *consumption.* Yet we sometimes find large families perish with it, whose parents had showed none of its signs. Persons of the scrofulous habit of body

5

are distinguished by the clear whiteness of the skin, the bright red of the cheeks, the narrowness of the chest, projecting scapulæ, or shoulder-blades, slenderness or deformity of the limbs and trunk, combined perhaps with swollen abdomen, and general want of bodily symmetry; there is a waxy yellowness round the mouth, the face indolent, and without energy. The eyes have a pearly, or blueish whiteness, with long silky lashes. In a second class of persons the complexion is dark, the skin harsh, and the habit indolent; the face has a swollen and pasty look, and all the functions of the body are sluggish and imperfect; the nervous energy is feeble, the feelings obtuse, and the moral and intellectual powers occupy a very low rank. But this terrible malady not unfrequently cuts off the most robust, and those who have the finest conformation both of body and mind.

Females more subject to Consumption than Males.

Woman has naturally a greater delicacy of constitution than the other sex, and is less able to bear up against the physical ills of life; but this fact is to be attributed very much to faulty habits of female life; to the wearing of corsets, by which the chest is cramped, and forced out of shape; and to exposure of the upper part of the chest; and to the habit of wearing insufficient protection for the feet, producing frequent derangement of the female system. They also breathe less pure air, from their habits keeping them more confined in the hot rooms and close air within doors, and impure air is a prominent cause of scrofula and consumption.

Anatomy of the Lungs.

These are contained in the *thorax or chest*, which is a cavity in the upper part of the trunk, larger below and smaller above. The lungs and heart occupy the principal part of this cavity, which extends from the collar bone and shoulders, as far down as the diaphragm, which separates the lungs from the cavity of the abdomen, or belly.

The diaphragm may be termed the floor of the lungs; it is an expansion of muscle and tendon, or substance like gristle, and is not flat, but is higher in the middle than at the edges, where it is attached to the lower and inside edges of the lowest ribs. The lungs, therefore, occupy nearly half of the whole inside space of the trunk of the body. The chest is divided in the middle, up and down, and from the breast bone to the back bone by a thin partition of membrane, called the mediastinum. On the right side is the right lung which has three lobes or parts; on the left side is the left lung, which has two lobes; and these nearly surround the heart. The heart and lungs are intimately united by blood-vessels; and the lungs, near their summits, by two air-tubes called *bronchi*, which are branches of the trachea, or windpipe. These blood-vessels and air-tubes are called the roots of the lung. The lungs are covered, and the inside of the chest is lined by a glistening and bluish white membrane called *the pleura;* the part which covers the lungs being reflected on, or joining the other, and the two surfaces being in contact.

Dr. Rush describes the lungs as made in the following manner: "An elastic air-tube, called the trachea or windpipe, opens into the upper portion of the throat, and consequently communicates with the mouth and back part of the nostrils — by a curious mechanism, in which the voice is mostly formed, called the larynx. This tube passes down the neck, enters the chest, and then forks into two divisions, called *bronchia* or *bronchi*, from a Greek word meaning the throat, one going to each lung. They then subdivide and go on branching again and again, becoming smaller and smaller, and less and less elastic, until they ultimately terminate in the minute vesicles, or air-cells, to which I have before alluded. These air-cells, with the air-tubes conducting to them, may be viewed as the framework of the lungs, and constitute the greater proportion of their substance. The cells, too, always contain more or less air ; it is to them that these organs owe their light and spongy character.

" The union of these cells is effected through the medium of a fine membrane, denominated cellular, which, though so abundant in many other structures of the body, is here very small in quantity, Everywhere upon these cells minute vessels are ramifying, to carry to them blood to be acted upon by the vital air they are continually receiving ; and to convey it back again in its course to the heart, after having undergone its mysterious aërial change.

" It appears to have been a grand principle of nature, in building up the beautiful and important organs of respiration, to provide that the greatest possible quantity of blood should be brought under the influence of the greatest possible amount of air. The number of air-cells almost exceeds calculation. They have been estimated in man at between one and two hundred millions, and as presenting a surface of two thousand square feet. They exercise, too, as may readily be conceived, the most important agency in the breathing function, since it is during the passage of the blood over their delicate coats, that the essential vital influnce is wrought upon this fluid.

A very tolerable idea of the appearance of the human *lungs* may be obtained by examining the *lights* of a swine. They should be inflated or blown up, by inserting a quill into the air-tubes.

Respiration or Breathing—how performed.

The general principle of the operation of breathing is as follows. — The lungs are contained in a shut cavity, which is closed in at the top and sides by the ribs, and other bones, the spaces between which are filled by the muscles, or red-flesh ; and is bounded underneath by the diaphragm. The lungs, with the heart, completely fill this cavity, their external surface covered by the pleura, being everywhere in contact with the pleural lining of the inside of the chest. But the contents of this cavity, as to space, can be greatly altered by the motion of the ribs and the rise and fall of the diaphragm, or floor of the lungs. When this space is diminished the lungs are compressed, and a portion of the air contained in them is expelled through the windpipe. This constitutes an *expiration*.

On the other hand, when this space is increased, the air rushes into the lungs, through the windpipe, in order to keep the lungs full, and prevent a partial vacuum. or void space. The diaphragm does the most of this work. The contraction of it pulls it down in the middle, as when we draw in a long breath or sigh, and thus enlarges the capacity of the chest, expands the lungs, and cause air to enter through the windpipe to fill them. This causes an *inspiration*.

When breathing quietly, this action is alone nearly sufficient to produce the requisite enlargement of the chest. In the movement of expiration, the diaphragm is quite passive; for, being in a state of relaxation, it is forced up by the liver and other abdominal organs under it, which are pressed in by the contraction of the abdominal muscles.

These last, then, are the main agency of the movement of expiration, diminishing the cavity of the chest by lifting its floor, and thus pushing out the air.

The action of the muscles of the abdomen, and those of the spaces between the ribs, also contribute to enlarge or diminish the capacity of the chest, by separating or drawing together its sides, and thus drawing in or forcing out the air. The action of the chest is thus seen to be much like a common pair of bellows; the lungs themselves being nearly passive, as far as drawing in and forcing out the air is concerned.

We have already seen that the lungs are not simply hollow organs, but are composed of an immense number of air-cells, whose duty it is to purify and renew the blood, by converting the dark red or venous kind into bright red or arterial. This is effected by its exposure to the action of the oxygen of the air in these minute cells.

The delicateness of the internal surface of the lungs will appear evident when we consider their lightness or trifling weight. Notwithstanding this great amount of surface, thirty times that of the whole surface of the body, as some have reckoned, the lungs, in the largest individual, weigh, at most, only a few pounds. The lining membrane is so delicately and finely constructed that it readily allows of the transmission of air through it, while the blood is retained in the appropriate vessels for that purpose. In health, there are seventy-two pulsations of the heart per minute, and the amount of blood sent to the lungs at each pulsation of the heart, in an adult individual of average weight, may be reckoned at about two ounces. According to this estimate, then, more than *twenty-five hogsheads* of blood are sent through the heart, and to the lungs, every day of the individual's life ; and, to purify this blood, more than twice the bulk of air must be inhaled.

When, therefore, we consider the important relations that exist between respiration and circulation, and the most intimate dependence of life at all times on these functions, numerous practical inferences are suggested to the mind. Thus it will be apparent that the *quality* of the air breathed must ever be an important consideration in regard to health. So, too, the quality of the blood, which is to be changed constantly by the wonderful action of the air upon it! What food shall we eat? What air shall we breathe? What exercise shall we take? These are important questions, and such as concern every individual, in proportion as health is the best of all earthly gifts. These considerations will be entered into more in detail in another part of the work. Nor shall we neglect the consideration of that most important point of all : — In

what way shall this vital fluid, the blood, be
purified and renovated when vitiated and cor-
rupted by hereditary taint, or by scrofulous,
tuberculous, cancerous, or syphilitic contamination. See page 47.

Tubercle the Foundation of Consumption.

The scrofulous diathesis, or habit of body,
as developed in *Tubercle* is now universally admitted to be the germ of true consumption.
The word is derived from the Latin *tuberculum*,
which means an excrescence, or tumor of some
part of the body. As here used we understand
a peculiar morbid product, or species of degen-
eration, depending upon imperfect nutrition and
impurity of the blood. Tubercles, themselves,
are composed of unorganized matter, or substance, deposited from the blood, of a yellowish
color, opaque, easily broken down, and of about
the firmness, and consistence of cheese. They
are most commonly deposited on the inside surface of the air-passages, and other mucous membranes, and very often among cellular tissue,
such as the air-cells of the lungs. Tubercles are
not necessarily of a round shape. Their form
depends upon the shape of the tissue in which
they are deposited. Wherever this begins it is
liable to increase by continual addition, and
fresh deposit. Hence when it begins in any soft
uniform substance, such as the brain, or cellular
membranes, the pressure being uniform on all
sides, it preserves a round form as it grows larger. *In the lungs* the round form is sometimes
real, sometimes *apparent* only. It is *real* when
the tubercular matter fills up, and therefore takes

the shape of the air-cells. So it is when a number of the seadjacent to each other coalesce, by the increase of deposit, and form one large round mass. But when, as is often the case, the tubercular matter is laid down in the smaller branches of the air-tubes, it assumes a cylindrical or pipe-stem shape. This we have many times ascertained by carefully following the branching of the air-tubes; but in the manner in which, after death, the lung is cut by the knife of the anatomist, we *see* merely cross-sections of these cylinders, and then the round form is *apparent* only. If the tubercular matter comes to fill one of these smaller air-tubes, and also all the air-cells to which that tube leads, then the morbid substance, when fully seen, represents a twig with a bunch of leaves at its extremity. Besides this true and undisputed kind of tubercle, the lungs are often more or less thickly studded with a number of small granules of firmer consistence, like cartilage, or gristle, semi-transparent, and of a blueish gray color. These are called *miliary tubercles*; and are formed in this way. The mucous membrane, lining the air-passages, secretes from the blood not only the matter of tubercle, but its own proper fluid; whence it happens that points of tubercle become enclosed in globules of gray tough semi-transparent mucus. *Miliary tubercles*, and the true cheesy kind, both occur in the same lungs, and in the same parts of the lung. One very seldom occurs without the other. They both belong, essentially, to *Pulmonary Consumption*.

The tubercular matter, then, being deposited on the inside of the small bronchial tubes, and of the air-cells, groups of these lying near each other, become united into one large mass; and generally there are tubercles of various sizes, from that of a grain of wheat, to a hen's egg, in the same diseased lung.

These large tubercular masses soften, and break down in the following way: — The *tissue* of the air-cells, and tubes, which originally separated these tubercles, still exists, though it ceases to be visible. At length, under the increasing pressure, or in some other way, it ulcerates, or suppurates; and the tubercular mass softens between its component parts, and thus breaks up. In the following way they are expelled from the body: They increase till the surrounding parts inflame. The inflammation thus excited, occurring in scrofulous persons, has the scrofulous character. The thin pus, or matter, which is thrown out, pervades, and loosens the tubercular deposit; a process of ulceration goes on in the surrounding textures; and, at last, the softened scrofulous matter is gradually coughed up and expectorated.

We have spoken of two kinds of tubercle: the common, or cheesy kind, and *miliary tubercle*. There is yet another form which the tubercular matter is apt to take. It sometimes is diffused uniformly over a considerable space; occupying all the cells, and interstitial portions of the part affected; and has no distinct boundary. The part looks as if liquid tubercle had been poured into it, and there had hardened. This has been called *tubercular infiltration*.

It is an important fact that tubercles are not deposited at random in the lungs. It is in the upper lobes, and in the upper and back part of those lobes, in nearly all cases, we find them when they are few. It is in the same part that they are largest, and most numerous, when they are scattered throughout the whole lung. It is here also that they first ripen and grow soft, and become ready for expulsion through the air-tubes and windpipe. Thus it is here that we find the most numerous, and the largest excavations, or cavities in the lungs. And the number and size of the tubercles, and the cavities formed by their discharge, gradually diminish from the top of the lungs downwards. This point helps distinguish the solidification of the lungs, arising from tubercle, from that of simple inflammation; as the latter prefers the lower lobes. The left lung is much more affected by tubercle than the right. Dr. Rush mentions that out of seventy-six cases in which the upper lobe was totally disorganized, on one side, fifty-six were of the left and only twenty of the right.

Tubercular matter, once deposited, remains for some time in the crude state, surrounded by healthy lung, and undergoes no change in quantity or consistence. But in most cases scrofulous inflammation begins around the tubercles, or in the walls of the included cells, and then the whole softens, as before mentioned, and this broken down substance is conveyed through the smaller air-tubes, into the bronchial tubes, and thence is coughed up through the windpipe, and expectorated. Of course, *a cavity*, or void space, which the tubercular deposit previously filled, is left. These cavities may be no larger than a bean, or large enough to hold a pint or more of liquid. Sometimes the whole of the upper lobe is converted into a bag of this kind. These large cavities are formed by the union of several smaller ones; and are often of very irregular shape, and divided as it were into chambers, by imperfect partitions, or by bands, which cross them in different directions. These large cavities are never met with in the lower lobes. Opening into such a large cavity there is always one, and may be several open air-tubes, which look as if they had been cut off just where they enter. We almost never find a blood-vessel thus open. The reason of which is that the air-tubes are filled up by the tubercular deposit, and when this breaks down they are destroyed with it ; the open mouth serving to carry the softened tubercle towards the windpipe and mouth. But a blood-vessel contains a coagulable fluid, and, when pressed upon by the tubercular substance, yields to the pressure, and the circulation of the blood being stopped, it coagulates, and seals up its vessel for some distance. Afterwards, when the cavity forms, the obliterated vessel is destroyed along with the portion of lung which originally contained it. In this way the erosion, or eating off, of any large vessel is prevented ; but it does occasionally take place, *though very rarely*, and is then immediately fatal. The popular idea that the " bursting of a blood-vessel " is the common and usual cause of bleeding at the lungs, or stomach, is quite incorrect. See *Pulmonary Hæmorrhage*. An oozing, or exhalation, of small quantities of blood may take place from the inner surface of the cavity, tinging the matter expectorated.

When the cavity is first formed, its inner surface is soft and ragged ; and if other softening tubercles are near, the cavity goes on enlarging, by two or more becoming one. This process is commonly called *ulceration ;* and a person who has a cavity of any size is said to have the lungs *ulcerated*. A cavity formed as above may remain stationary, if there are no other tubercles near.   Its inner surface then becomes smoother, and a lining membrane forms upon it.   This may or may not secrete pus, or matter.   Generally the substance of the lungs round such a cavity becomes condensed and solidified.

When cavities occur singly, without other tubercles, or cavities, they may be completely emptied of the tubercular matter ; may gradually contract ; and ultimately becomes quite obliterated.   Their inner surface become converted in such a case into a substance resembling cartilage, or gristle, and presents the appearance of a cicatrix, or scar.   The process which has gone on is a process of recovery ; and this would be complete if no fresh deposit of tubercle took place.   Such cases, however, when left to nature or usual modes of treatment, usually have tubercles multiply, until at length their effects terminate life.   Another way in which tubercles, when limited in number and extent, may cease to cause disease, is as follows : — The more watery parts of the morbid secretion may be absorbed, and the earthy salts it contains may concrete, and the whole be converted into a shrivelled, hard, chalky lump, which sometimes is coughed up; or, in favorable cases remains for years in the lung, an almost harmless body.   The expectoration of such concretions marks the chronic nature of the case.   The editor has for some time treated a lady, who, though not robust, enjoys very tolerable health, and who has, at intervals, coughed up small hard lumps, and even branching fragments, evidently moulded in the smaller air-tubes.   They resemble white coral, and chemical analysis shows them to consist of phosphate and carbonate of lime.   Dr. Rush, also, relates such a case. Cures of this kind are much more common, now, under improved treatment.

And this now brings up more prominently the whole question of the *curability of consumption*, after cavities have formed. We shall first mention well-known opinions and instances.

It has been a prevalent opinion that, when once a cavity has formed in the lungs, it is incapable of cicatrization; but the author has met with several instances in which this change was unequivocally accomplished; and the details of the case of an eminent medical practitioner of Philadelphia, Dr. Parrish, have been published, in whose lungs there were marked evidences of cicatrization. This occurred after he had long suffered under symptoms of phthisis, and exposed himself to a regimen as will be mentioned hereafter. M. Boudet affirmed, before the *Academie Royale des Sciences*, of Paris, that, in 197 cases taken indiscriminately, he found ten examples of cavity completely cicatrized, without any trace of recent tubercles; and eight examples of complete or partial cure of cavity, coinciding with recent tubercles; and, he concludes that recovery is possible at any period of pulmonary consumption. His researches would seem to show that tubercles of the lungs are common when no suspicion is entertained of their existence. Of the 197 persons who died in the hospitals of various diseases, or were killed suddenly by accidents, from the age of two to that of fifteen, tubercles were present in three-fourths of the cases; and in those between fifteen and seventy-six, no fewer than six-sevenths of the bodies exhibited tubercular deposits in the lungs; — these results are explained by the facility with which tubercles undergo a change in their intimate constitution, by which they become not incompatible with a state of health; these modes of cure consist of sequestration, or separation, induration, and absorption.

Careful observers are constantly discovering evidences, on dissection, of healed cavities. In seventy-three bodies examined by Dr. Rush, he found cicatrices, or scars of healed cavities, in twenty eight. In twelve, hardened spots alone existed; and in sixteen, calcareous concretions, or dry and harmless tubercle, alone were present. The editor of this volume has not been less fortunate. In quite a number of cases cured by the remédies hereafter mentioned in this book (see page 54), death afterwards took place from accident, or from some other disease; and unmistakable evidence was found of healed cavities. In one case a cure took place after the whole of the upper lobe of the left lung was destroyed, and two healed places were found in the upper lobe of the right lung.

In 151 bodies examined at the hospital of Larriboisier, at Paris, while the editor was attending there, seventy-six, or more than one-half, showed marks of healed cavities.

Dr. Hughes arrives at the startling conclusion, from his own observations, and those of Drs. Rogée and Boudet, that the cure of tubercle has occurred in from one third to one half of all who die after forty. " So deeply rooted, however," he adds, " is the opinion of the necessarily fatal nature of this disease, that, simply because recovery has taken place in certain cases, medical men have rather mistrusted their own diagnosis, than ventured to oppose a dogma of universal belief."

Dr. Cullen, who was a practitioner of great

celebrity and experience in the last century, regarded that the prospect for recovery in consumption is for the most part unfavorable; and
that of those affected with it, the greater number
die. " But there are always," says he, " many
of them who recover entirely, after having been
in very unpromising circumstances." He published the following aphorisms as the result of his
observations : —

" 1. A consumption arising from pulmonary
hæmorrhage (bleeding) is more frequently recovered from than when arising from tubercles.

" 2. A pulmonary hæmorrhage not only is
not always followed by a consumption, but even
when followed by an ulceration, this latter is
sometimes attended with little of hectic, and frequently admits of being soon healed. Even when
the hæmorrhage and ulceration have appeared
to be repeated, there are instances of persons recovering entirely after several such repetitions.

" 3. A consumption from tubercles has, I think,
been recovered from; but it is the most dangerous of all; and when arising from an hereditary
taint, is almost certainly fatal.

" 4. The danger of consumption, from whatever cause it may have arisen, is most certainly
to be judged of by the degree to which the hectic and its consequences have arrived. From a
certain degree of emaciation, debility, profuse
sweating, and diarrhœa, no person recovers;
but this degree constitutes, under the use of
remedies here advised, the very last stage of this
disease.

Various other authors have given accounts of
cases in which it was found after death there were
cicatrizations, or scars, in the substance of the
pulmonary tissue, proving, to a most positive

demonstration that ulceration had at some period existed. Every medical man, too, who has had large experience in treating this disease, must have found cases in which all the more prominent signs of consumption existed; but in which recoveries took place even contrary to his own expectations.

Considering, then, how alarming its ravages have become, and at the same time the pernicious character of the physiological and hygienic habits of the community, may we not confidently hope that, by the promulgation of correct principles of health and medical treatment, much may yet be done to stay the march of this scourge of our climate. Beyond a doubt many may be saved who, under the present misunderstanding of the laws of health and the prevalent system of pernicious drugs, now sink into the consumptive's grave.

Before, however, leaving the subject of tubercle, we must mention a beautiful provision of nature for preventing a cavity from opening, or ulcerating its way through the side of the lung, and thus allowing the air and diseased matter from escaping into the cavity of the pleura, or space between the outside of the lung and the inside of the chest; which accident would be almost necessarily fatal. When the tubercles are numerous, and lie near the surface of the lung, this causes an adhesive inflammation to arise between the membrane called the pleura, where it covers the lung just over the diseased spot, and the inside of the chest, just opposite, so that the two surfaces are united, and grow permanently together. This is called a pleuritic adhesion. The top and back parts of the upper lobes, in very many, or nearly all confirmed consumptives, are

6

united, in this way, to the inside walls of the chest.

TUBERCLES FOUND IN DIFFERENT ORGANS.

We have before stated that *tubercle* and *consumption* are both diseases of the whole system. Tubercles are found *most frequently* in the lungs and glands of the throat; next in the glands of the neck, and those of the abdomen; next in frequency in order come the spleen, pleura, liver, intestines, brain, and kidneys. In all cases of tubercles of the lungs, they are also to be found in one or more organs besides, usually several. In view of this fact we cannot too strongly impress it upon the reader that the disease itself begins in the blood, and continues to be engendered there; and that the plain inference from all this is that no cure, or even permanent improvement can be expected except under the action of such remedies as purify and renovate the blood. See pages 47 to 54.

Dr. Rush gives similar testimony in regard to tubercles, he says, "There is not an organ of the body but is capable, as well in its substance, as in its covering, of producing tubercles of some kind or other; and occasionally of almost every kind at the same time." As an instance of this may be cited the air-passages that lead into the lungs as very liable to become affected. The mucous membrance of the larynx, and windpipe ulcerates, and giving rise to hoarseness, and loss of voice, the disease is sometimes called *laryngial consumption.* But this is not a disease of itself, as the *scrofulous* ulceration only occurs when there are tubercles in the lungs. It is quite probable that the mucous glands of the throat are incited to ulcerate by the passage over them of the expectoration of tubercular matter. Nearly one consumptive person in three has ulceration of the windpipe. And (*except syphilis*) ulceration of the larynx and windpipe is peculiar to pulmonary consumption.

The glands of the bowels are very commonly much affected by turbercle. And, this in the latter stages of consumption, gives rise to the consumptive diarrhœa. In negroes, tubercle very commonly attacks the glands of the mesentery and bowels, in preference to the lungs; and is known, at the South, as *negro's consumption.*

The liver, in consumption, is liable to undergo a remarkable change which, is almost peculiar to this disease. It enlarges, and becomes full of a fatty substance, which greases the hands and knife of the anatomist, and paper, when placed upon it. The whole liver loses its natural dark red tint, and assumes a pale yellowish color, and becomes softened. The enlargement is frequently very great, so that instead of weighing four pounds, it weighs seven or eight. This state of the liver is more likely to occur in persons who drink freely of malt and other spirituous liquors.

Different Forms of Consumption.

It has already been stated that all ages are subject to this disease; and since morbid anatomy has been more studied, children have been found more subject to scrofulous and tuberculous complaints than was formerly supposed.

Dr. Guersent, of Paris, one of the physicians to the Hospital des Enfans Malades — an institution appropriated to the treatment of patients between the ages of one and sixteen years — states, as the results of his observations, that five-sixths of those who died in that establishment are more or less tuberculous. The symptoms of consumption, in cases of children, are less strongly marked than in adults; and it is to this fact that we must attribute the former belief, that tuberculous disease can rarely happen in infants. The hectic fever, in children, is less perfectly formed than in adults. The perspirations are less abundant, and cough occurs frequently in paroxysms; resembling those of pertussis, or hooping-cough.

In many cases there is little or no expectoration, for the reason that the matter is often swallowed; so that it attracts no attention. We may usually detect, however, the tuberculous look of the child, like scrofula; the rapid pulse and breathing, the frequent cough, and the gradually increasing emaciation.

There is often much derangement of the digestive organs. The bowels are swelled, sometimes constipated, sometimes the reverse; the evacuations pale, and of unnatural color. The mesenteric glands (those contiguous to the bowels) are more frequently tuberculous in children than in adults, constituting a variety known as *Tabes Mesenterica.*

In the varieties of pulmonary consumption, we make, then, the following : —

1. That which affects infants and children; an obscure form of the disease, and one which it is sometimes impossible to detect. In adults we observe three prominent kinds; the *chronic* variety, in which the symptoms are more marked, and distinct in character; and which many pass into the more acute form, or be merged into it.

2. The acute variety, which happens oftenest in young persons, especially young females; runs its course frequently with great rapidity, causing

death in one, two, or three months; but which
may also precede, or follow, the latent and
chronic varieties of the disease.

3. That which is *latent;* and which is often
very difficult, and sometimes impossible to detect.
The chronic form is so called from its duration;
though all forms are more or less chronic. Slow
cases of this kind may last ten, twenty, or even
forty years. It is therefore the more usual form
in older persons; and is apt to occur even after
forty years of age.

Acute, Rapid, or Quick Consumption.

The duration is more commonly from six
months to two years; but what is popularly
termed quick consumption, runs its course in two
or three months, or even less. This has sometimes been called *galloping consumption.* It is
more apt to assume this rapid form, when developed by some other disease, as measles, small-
pox, or inflammation of the lungs.

The tuberculous habit of body, the morbid
matter in the blood, or even small tubercles
already formed, may, and often do remain long
quiescent, until some accident, such as exposure
to cold, great exertion, an attack of catarrh, or
hæmorrhage, or inflammation of the lungs, serves
to develop the malady, which has for a long time
been doing its silent work. This form is more
common in young persons, particularly in young
females; and more often in those who inherit
the scrofulous habit. It has been often observed
as occurring from suppressed or deranged menstruation. Such persons seem as frail as autumn
flowers, and fail almost as readily. They are

often peacefully sinking into the arms of death before they suspect their danger.

It will hence be seen that the *prevention of consumption* is a very important point for those inclined to scrofula, as well as others.

Thus, briefly, have we noticed the different varieties of *pulmonary consumption.* But we must remember that we never find two cases exactly alike, and that oftentimes one kind merges into another. The *latent* kind becomes active, or the *chronic* may suddenly become acute. The *acute* variety, too, may change, for the time. into the *chronic.*

CAUSES OF CONSUMPTION.

These come under two classes · — first, the *remote,* or *predisposing* causes; secondly, the *exciting* causes, or those which call the predisposition into action.

Hereditary transmission is the chief *remote* cause. It is as certain that children inherit the diseases of their parents, as that they resemble them in features and character.

In proportion to the development of the tuberculous disease in father, or mother, will be the disposition to the same affection in the child.

Any disease, or any circumstances which can impair the health of one or both parents, materially influences the health of the child yet unborn; thus many persons acquire a predisposition to consumption from their parents, although the latter may attain an advanced age without evincing any symptoms of pulmonary disorder.

Indigestion, many cutaneous affections, syphilis, anxiety, grief, and the depressing passions, intemperance or irregular mode of life in the mother, with insufficiency of proper nourishment during pregnancy, are all capable of inducing a scrofulous habit, and, as a consequence, a predisposition to consumption : that which was bad general health iu one generation, is frequently converted into tuberculous disease in the succeeding one.

A peculiar formation of body, a distorted spine, narrow chest, and high shoulders, must also be considered a remote cause ; and every pulmonary affection occurring in persons thus shaped, should always be looked upon with suspicion, even in the absence of hereditary predisposition, or more decided exciting cause.

The question will probably occur to many — can a child, born of healthy parents, free from scrofulous taint ; can he, in after life, become affected without tuberculous disease ? — that is, can tubercle *originate* in him ? It can. By the combination of many circumstances, which will be noticed under the head of Exciting Causes, a morbid state of the system is established which induces and favors the deposit of tuberculous matter ; and by the continuance of such causes, he may fall a victim to consumption. If a child born of healthy parents be insufficiently or improperly fed, or nursed by an unhealthy nurse or be confined in a dark, ill-ventilated room, wallowing in dirt and uncleanliness, tuberculous disease will, in all probability, be engendered, and the scrofulous habit, with its symptoms, be acquired.

The peculiarities of frame and appearance, which mark this habit, or diathesis, have already been described. They are characteristic of a dormant liability to consumption. By such persons its exciting causes should be carefully guarded against. And suitable medicines taken to eradicate this lurking evil from the system.

To recapitulate : — Tubercle is the seed of the disease ; it may be hereditary ; it may be acquired ; an individual may possess undoubted signs of its existence ; he may have the scrofulous diathesis strongly marked ; he may have lost brothers and sisters, father and mother, by the disease, and yet he, by preventing the germination of this seed, may escape. It, therefore, behooves such an one to avoid the thousand circumstances which may act as a hot-bed in ripening this seed ; some of which we now proceed to notice.

EXCITING CAUSES. — Many exciting causes, when acting together in early youth, as improper diet, impure air, deficient exercise, insufficient clothing, and the absence of cleanliness, readily become a *remote* cause, capable of engendering the disease. *Food* which is not sufficiently nutricious, and food that is too rich and stimulating, are alike hurtful : the former does not furnish an adequate supply of nutriment to support·the body in health and strength ; the latter excites and irritates the digestive organs, and produces indigestion, — one of the most frequent and active agents in exciting consumption.

Eating Pork, or Swine's Flesh,

in any form, is one of the most prolific causes of this disease. It is perhaps not generally known that all swine are more or less scrofulous, or tubuculous, and would die of such diseases if left to die a natural death. The very name of a hog

also means scrofula, in some languages. It was
not without reason that the Jewish lawgiver prohibited pork-eating; and if pork was not fit to
eat *then*, it is not fit to eat *now*. The Jews, to
this day, abstain from it ; and are nearly free from
consumption and scrofula. Our New England
farmers, and in fact that class of men all over
the Union, from their out-of-doors occupation,
and from their general good habits, ought to be
nearly exempt from these diseases. Yet they
perish, and their families are ravaged by this destroyer; and they owe it to pork-eating, and
very little else.

Bad Air as a Cause of Consumption and Scrofula.

It will be almost invariably found that a truly
scrofulous disease is caused wholy, or in part, by
sleeping, working, or other exposure to a vitiated and corrupt atmosphere. Or, at least, that
no person so exposed, for any length of time,
will escape such contamination. Nor is it necessary that there should have been a long stay in
bad air to cause it. A few hours each day is
sufficient; and thus persons may live in a healthy
place, pass much time in the open air, and yet
become scrofulous, or tuberculous, because of
sleeping in a confined place, where the air is not
renewed. In Europe this is seen in the case of
many shepherds, who perhaps wrongly refer their
complaints to exposure to the weather. But it
frequently escapes attention that they pass the
night in a close-shut hut, or tent, which they
move from place to place ; and which completely
protects them from dampness. Six or eight
hours a day passed in a close and exhausted air

is the true cause of their disease. The bad habit of sleeping with the head under the bed-clothes has the same effect, as the editor has often witnessed in the negro's consumption at the South: this being a very prevalent habit with them.

The following is a striking example of the fatal effects of bad air.

At a distance of ten miles from Amiens, France, is the village of Meaux. It is in a vast plain, slightly elevated above the neighboring vales. Seventy years ago the houses were built of clay, without windows, lighted by a pane or two of glass in the walls. There were no floors, save the earth. The inhabitants were engaged in weaving. A few holes in the walls, which were closed at will by a plank, scarcely permitted the air and light to penetrate into the workshops. Dampness was thought necessary to keep the threads fresh. Nearly all these people were seized with scrofula, or consumption, and many families thus became extinct. After a time a fire destroyed about one third of this village; when the houses were rebuilt in a more salubrious manner, and by degrees scrofula became less common, and disappeared from that part. Twenty years later another third of the village was consumed ; the same amelioration in building had a like effect as to scrofula, and the disease is now confined to the inhabitants of the older houses, which retain the old faults.

Dr. Rush has affirmed that foul air and deficient ventilation is more productive of consumption than all other causes.

In short, pure *air*, and plenty of it, is the basis of health: if impure in quality, it irritates the delicate structure of the lungs, and impedes respiration : when fresh air is insufficient in quantity, it is unable to assimilate the chyle, or nutritious element of food, during its circulation through the lungs. A prolific source of disease is found in the practice, too frequently unavoidable, of many persons sleeping in the same chamber.

The want of light, as a cause of consumption. Light is one of the most important agents in the growth and well-being of both vegetable, and animal kingdom. Plants growing in cellars, celery, growing with its stalks protected from the light, are instances of the former. In the animal kingdom we may cite tadpoles, which afterwards should become frogs; if these are kept from the light, their metamorphosis into air-breathing animals does not take place, and they only remain large tadpoles. In large hospitals it is a familiar fact that the wards and rooms on the south or sunny side are much more favorable to recoveries than those on the north side. This has been particularly observed in London, where there is always a deficiency of sunlight.

In accordance with these facts, we find that consumption goes most among those who live in buildings or rooms where the sun's rays never come.

Deficient Exercise a Cause of Consumption.

A *sedentary life* in youth arrests the growth and proper development of the body; in mature age, it impedes or disorders every function. Statistics clearly prove that the disease is more prevalent in cities and manufacturing towns than in the rural districts, where the population has plenty of exercise in the open air; and that it is more prevalent amongst clerks, tailors, shoe-

makers, and watchmakers, than it is amongst
sailors, carpenters, and others whose occupation is
active. The want of exercise is an exciting
cause of consumption, which is constantly over-
looked or misapprehended even by the most anxious parents; under the dread of *fatiguing* a
delicate child, they restrict him or her to unnatural and unhealthy quietude; and this incorrect
idea is zealously carried out at fashionable, and
too frequently *finishing*, boarding schools, where
every movement is regulated by rule.

Children should be children, and abundant
exercise in the open air is one of the essentials
for their healthy development. Their noisy
plays, therefore, should not be discouraged, but
rather promoted, in the remembrance that, unless
they are healthy *children*, they will not be
healthy *men and women*.

Excessive Mental or Physical Labor

may sap the vital functions, and cause consumption.

Mental labor in excess may exhaust the nervous energy of the system, and thus indirectly
help other causes ; or, it may directly be a cause
by depriving the body of sufficient pure air and
exercise.

Over-exertion of the Lungs.

A proper degree of exercise of these, and
other vocal organs, is conducive to health. Such
exercise calls the muscles engaged in respiration
into free play, and aids in producing full development and expansion of the chest and lungs.

But violent and long continued efforts in play-
ing wind-instruments, or in singing, or public

speaking, are necessarily prejudicial to those who are predisposed to disease of the lungs. Reading aloud, however, from day to day, singing, and frequent public speaking, when not carried to the extent of much fatigue, are excellent means for invigorating the lungs. We must here, as in other cases, distinguish between the use and abuse of valuable means.

Improper Clothing.

Ladies suffer more than the other sex from this cause. Generally, too great an amount is worn. Then, to attend a ball, party, or the like, a great change is made. Their stout stockings and every-day shoes are changed for less substantial ones. The same is true of the dress. By these improper changes, cold is often taken that lasts for weeks and months, and in many cases the foundation for consumption is thus laid. Great changes in regard to dress should never be made suddenly, even with persons in health. Dr. Rush was so impressed with the importance of this rule, that he used to say, quaintly, that all who wear flannel should, in mid-summer, put it off one day and resume it the next. We repeat it, there is generally too much clothing worn; but all changes to a less amount should be made in connection with cool or cold bathing, and should be as gradual as possible. Many a cold is taken, when if, at the time of making the change, a cold bath, a sitz-bath, or even a sponge-bath had been taken, followed by exercise, the person would have been safe.

The practice of wearing low-neck dresses is often injurious. About the waist there is worn a half dozen or more thicknesses, while about the top of the lungs there are none at all. Usually, perhaps, the same person will have the neck warmly clad. Such changes are bad, and persons are often injured by them. The skin is a breathing organ, almost as much as the lungs; hence, clothing should never be so tight as to prevent free access of air to the surface.

A constant cause of disease in females is tight-lacing, by which the lungs and abdomen are unnaturally compressed. The corset is a most barbarous piece of armor, which frequently destroys the beautiful symmetry of the natural form, and exhibits, instead, artificial deformity. Imagine the Medicean Venus reduced to a genteel waist by a pair of stays!

Calomel, or Mercury, a Cause of Consumption.

Dr. Rush observes that this mineral quicksilver, when used so as to affect the system, in usual forms of calomel, blue-pill, mercury, and chalk, and a variety of other forms, is, without doubt, capable of inducing tuberculous disease. He considers that in persons of a delicate, or scrofulous constitution, if used at all, its use requires the greatest caution and circumspection. If mercury is capable of producing dropsy, enlargement of most of the glands of the body, sloughing, and ulceration of the cheeks, gums, and throat, mercurial fevers, diseases of the skin, tremors, and palsy, mecurial wasting of the bowels, and decay and disease of the teeth, and bones, — as not only Dr. Rush, but also Sir Astley Cooper, and many other eminent men bear witness, — if mercury may cause all these evils, which it does, with a mighty host of others, which not fifty pages of this book could describe, we may easily believe in the mercurial hectic described by Travers, and known *by irritable circulation, extreme pallor and emaciation, and acute and rapid hectic, and an almost invariable termination in consumption.*

Experiments made in France, on dogs, in which mercury was introduced into the lungs through the air-tubes, and into the cells of the long bones, showed tubercles, with a globule of mercury in the centre as the result.

"In the year 1810," says Dr. Sweetser, " a large quantity of quicksilver was taken from the wreck of a Spanish vessel, on board the English ship Triumph, of seventy-four guns, and the boxes principally stowed in the bread-room. Many of the bladders in which it was confined — owing to the heat of the weather, and to having been wetted during their removal — soon rotted, and several tons of the mercury were diffused through the ship, mixing with the bread, and more or less with the other provisions. The consequence was, that very many of the officers and crew experienced severe salivations, and other deleterious effects, from the mercury that was taken into their systems, two dying from its influence ; and that nearly all the live stock, as well as cats, mice, a dog, and even a canary bird, died.

"But how did it affect the lungs? The account informs us that the mercury was very deleterious to those having any tendency to pulmonic affections; that three men, who had previously manifested no indisposition, died of pulmonary consumption ; and that one man who had before suffered from lung-fever, but was quite cured, and another who had had no pulmonary complaint, were left behind at Gibraltar, with confirmed *consumption.*"

Hard Water may Cause Consumption.

Experience proves that the constant use of hard water, in connection with other bad habits, is decidedly injurious. There is no doubt that the use of water holding in solution considerable amounts of mineral matter, and more especially *dissolved limestone*, in limestone sections, is often a cause of *scrofula, tubercle, and consumption.* But some may ask, what then shall we do for water? We answer, that *filtered rain-water, from suitable cisterns*, is the very best, and an ample source of supply; and, as a rule, water that is too hard to wash with *is not fit to drink.*

Suppression of Habitual Discharges.

A sudden suppression of the monthly flow, as by taking cold, has been known to develop consumption. A sudden check of perspiration from exposure to cold; the checking of long established skin eruptions by stimulating washes and ointments; and the healing up of old sores and fistulous openings may also develop this disease, more particularly where there is a predisposition to it.

Sexual Abuses Cause Consumption.

Solitary vice is sufficient, as testified by the physicians of our insane asylums, to cause idiocy, the most lamentable of all the forms of insanity, oftener than all other causes put together. Their testimony is equally united that it frequently also causes consumption, preceded by a great amount of depression of spirits, melancholy, dissatisfaction with life, and waste of strength and energy, of both body and mind, to a deplorable extent. Both sexes are exposed, and become victims to this unseemly practice. Persons interested in this subject should read the article page 241.

Connubial excesses, also, often develop this disease ; as may be inferred from the fact that we often see the newly-married pass rapidly into decline.

Many a young man, too, in great cities, those hot-beds of pollution and vice, has often, by a career of licentiousness, been brought early to a consumptive's grave. If a young man contracts that most loathsome of all diseases, which arises from licentious habits ; and if, besides this he gets, as is generally the case, a course of medical treatment by which the system is saturated with mercury, iodine, and like medicines, he will be fortunate if he does not pass into a rapid consumption, or some other decline equally fatal.

Tobacco a Cause of Consumption.

From the habitual use of tobacco, in either of its forms — of snuff, cud, or cigar — says Dr. Rush, " the following symptoms may arise : A sense of weakness, sinking or pain at the pit of the stomach ; dizziness or pain in the head ; occasional dulness or temporary loss of sight ; paleness and sallowness of the countenance ; and sometimes swelling of the feet ; enfeebled state of the voluntary muscles, — manifesting itself sometimes by tremulousness, weakness, squeaking, and hoarseness of the voice, — rarely a loss of voice ; disturbed sleep ; starting from early slumbers with a sense of suffocation, or feeling of alarm ; incubus or nightmare ; epileptic or convulsive fits ; confusion or weakness of the mental faculties ; peevishness and irritability of temper ; instability of purpose ; seasons of great depression of the spirits ; long fits of unbroken melancholy and despondency; and, in some cases, entire and permanent mental derangement."

" Tobacco," says Dr. Woodward, " is a powerful narcotic agent, and its use is very deleterious to the nervous system, producing tremors, vertigo, faintness, palpitations of the heart, derangement of the stomach, and other serious diseases." Judging from such authority, we may readily believe the statement of the great Dr. Rush, that he once lost a young man, seventeen years of age, of a pulmonary consumption, whose disorder was brought on by the intemperate use of cigars.

Miscellaneous Causes of Consumption.

Personal cleanliness is a duty we owe to ourselves and to those with whom we associate ; it is a means of preserving health within the reach of all, and its importance will be admitted when we consider that the skin is constantly producing perspiration and unctuous matters, which readily mix with the dust and fine particles floating in the air, and which, if allowed to collect and remain on the surface of the body, form a coating that closes up the pores of the skin, prevent its healthy action, and give to disease another ally.

Intemperance in the use of spirituous and fermented liquors is one of the most prolific causes of consumption : when acting, as too frequently happens, in conjunction with bad, innutritious diet and insufficient clothing, whereby the body is excited and stimulated, not strengthened and protected, habitual intemperance is capable of becoming a remote cause, or the originator, of tubercles, as well as the ever-ready agent to hasten their development, should they already exist. The blanched, emaciated countenance of the dram-drinker faithfully corresponds with the diseased condition of his internal organs; and it may occur that an attack of that dreadful malady, *delirium tremens*, gives more decided evidence of the mischief and destruction effected on the nervous system. The dire effects of this debasing habit are not confined, unfortunately, to the drunkard himself; his progeny suffer, perhaps, in a still greater degree, and the frequency of tuberculous disease in the children of dissipated parents is a fact which can be confirmed by every physician of experience.

Change of temperature directly affects the respiratory organs, and conveys an exciting cause to the very seat of tubercle; we, therefore, find consumption most general and most fatal in climates that are subject to sudden alternations from heat to cold. America and Great Britain rank the first in this unenviable position. In those climates where the atmosphere is uniform, whether it be cold or hot, as in Russia and the West Indies, consumption is comparatively rare; whilst in America it carries off nearly one-third of the inhabitants; in England, about one-fourth; in Paris, about one-fifth; and in Vienna, one-sixth. As well as by those rapid climatorial variations which are native to our soil, the disease is nurtured by our own carelessness. This carelessness is directed rather to the effect than to the cause, for we constantly meet with persons who dread " catching cold," and use every precaution to avoid doing so, and yet they take no heed of the cold when it is " caught." The man who will not have his hair cut on an inclement day, lest he " take cold," will, nevertheless, allow a cold and a cough to distress him for weeks, without adopting any effectual means of removing it.

We do not remember having read a more forcible admonition on the necessity of attending to " a slight cold," than that written by the author of " The Diary of a late Physician." The value of the advice, and the vigor of the language, will be an adequate excuse for the extract : " Let not those complain of being bitten by a reptile, which they have cherished to maturity in their own bosoms, when they might have crushed it in the egg. Now, if we call a slight cold 'the egg,' and pleurisy, inflammation of the lungs, asthma, *consumption*, the venomous reptile; the matter will be no more than correctly figured. There are many ways in which this 'egg' may be deposited and hatched. Going suddenly, slightly clad, from a heated to a cold atmosphere, especially if you can contrive to be in a state of perspiration — sitting or standing in a draught, however slight — is the breath of death, reader, and laden with the vapors of the grave. Lying in damp beds, for there his cold arms shall embrace you ; continuing in wet clothing, and neglecting wet feet ; these, and a hundred others, are some of the ways in which you may slowly, imperceptibly, but surely cherish the creature, that shall at last creep inextricably inwards, and lie coiled about your vitals. Once more, again, we would say, *attend* to this, all ye who think it a small matter to neglect a *slight cold*."

Mental Emotion and the Passions,

Especially those which are depressing, exert a decided influence in arousing tubercles from their lair. The effect of mental affliction instantly overthrows the whole economy of the system ; an agonizing sense of oppression and tightness is experienced in the neighborhood of the heart and lungs, accompanied with a dreadful feeling of impending suffocation. If the sorrow be unremoved, if the heart be uncheered by hope, and this disturbance continues, the health sinks under the oppression, and the mind falls into despondency. In the downfall of long cherished hopes ; in the bereavement of a loved parent or friend ; in disappointed ambition ; in the reverse of fortune ; in slighted affection ; in fact, by all that " maketh the heart sick," — affliction of mind is a constant " worm i' th' bud," that preys on the health, and accelerates the progress of consumption.

A frequent exciting cause of consumption in young persons may be traced to a deep and settled despondency, consequent on a separation from the happy scenes and associations of home. This has been termed *home sickness* — "the piercing anguish hid in gentle heart" — the *heimwehr* of the Germans, the *maladie du pays* of the French. Whenever the sufferer from such a cause be of frail or delicate constitution, the danger will be greatly enhanced.

Intense application to study, which involves loss of sensorial power and exhaustion of the nervous system, together with sedentary habits, imperfect digestion, and constipation, is another mode in which the mental powers affect the health. One, from among the many victims of consumption hastened to an untimely end by severe mental application, was Kirke White — he who, whilst in the grasp of the destroyer, sang, —

" Gently, most gently, on thy victim's head,
Consumption, lay thine hand! Let me decay
Like the expiring lamp."

Rapid Growth

is, in many instances, the harbinger of this disease, as it is always attended by debility in consequence of inadequate nutrition : the progress of development in the frame being more rapid than the elimination of the required nourishment, the body grows without being matured, almost without being perfected. Richerand relates a case of this kind that terminated fatally, the individual having grown more than an English foot in a year.

Several *occupations* which produce mechanical irritation of the lungs, greatly quicken the development of tubercles : this mechanical irritation is excited by inhaling an atmosphere loaded with minute particles of dust or powders, as happens to sawyers, millers, starch-makers, flax-dressers, weavers, feather-dressers, and artizans similarly engaged. These employments, however, are harmless when compared with others in which the dust is of a deleterious nature, as it is in the manufacture of cutlery and the grinding of metals. The mortality amongst needle, edge-tool, and gun-barrel grinders, is excessive ; and Dr. Johnstone, of Worcester, informs us that the former seldom live to be forty. Mr. Thackerah gives a similar account of the early fatality of such employments in Sheffield, England, where the disease, so induced, is known amongst grinders by the name of "pointers' cough," or "grinders' rot."

Sedentary employments, and confinement in a particular position, are most injurious to those who have any predisposition to the disease : literary men, lawyers, artists, clerks, watchmakers, jewellers, tailors, shoemakers, and others similarly engaged, add more than their proportionate quota to the lists of mortality from consumptio-

Many diseases, especially those which affect the pulmonary organs, have a peculiar tendency to excite consumption : catarrh, bronchitis, and inflammation of the lungs, frequently give an impulse to the more serious and fatal malady. Fever, when occurring in a person of tuberculous constitution, acts in like manner. The eruptive fevers, as measles, small-pox, scarlet fever, frequently induce some subsequent disorder of the system, and in many instances that disorder is consumption. Nervous debility, produced by irregularity and excess ; indigestion, which implies deficient nutrition and constant irritation of the whole body, are never-failing causes ; worms, or anything capable of exciting habitual irritation in any part of the alimentary canal, readily induce a sympathetic action in the lungs. The tendency of syphilis to produce consumption has been noticed by almost every writer, from the time of Bennet (1654). Certain profuse discharges, as long continued diarrhoea, diabetes, menorrhagia, fluor albus, bleeding piles, &c., may, with sufficient reason, be included amongst the exciting causes.

The imprudent practice of young and delicate mothers suckling their children, as some do, for twelve or fourteen — nay, some eighteen months, or two years, is most reprehensible, and dangerous, alike to themselves and to their offspring.

It must not be supposed that these exciting causes act injuriously in every case, or that one alone is always sufficient to foster the disease ; but we may be assured that whatever tends to debilitate the constitution, whatever interferes with the proper nutrition of the frame, and whatever depresses the vital powers, will always accelerate and favor the production of tuberculous disease.

The opinion at one time prevailed that consumption was contagious; but the experience of modern physicians goes far to prove that it cannot be so propagated; it is, nevertheless, highly imprudent for a healthy person to occupy the same bed, or to sleep in the same chamber, with a consumptive patient.

Pulmonary Hæmorrhage. Bleeding at the Lungs.

This is often one of the earliest symptoms of consumption, and associated with it in many cases. In ninety-eight cases of consumption noticed by the editor of this book it occurred in seventy-seven, or about three-quarters: and this is a fair proportion in many hundred others. It is therefore somewhat an alarming symptom, or disorder; being frequently the first warning of incipient decline. Many persons, however, outgrow such attacks; or are undeniably cured by proper remedies.

Pulmonary Hæmorrhage is not necessarily. though usually, associated with tuberculous disease. It sometimes occurs as the result of disease of the heart, though such cases are rare. It is *more* common as *vicarious*, or *instead of* the monthly flow of females. Dr. Rush relates a case of a woman who menstruated quite regularly in this way for forty-two years. It rarely lasted more than one day and did not impair her health. Such cases, however, should be remedied by appropriate medicine adapted to regulate and restore the monthly periods. See page 47.

It has already been stated, page 76, that hæmorrhage from the lungs very rarely arises from the sudden giving way, or "bursting of a blood vessel." It arises simply from exhalation, or oozing of the blood, from the mucous membrane which lines the air-cells, and air-tubes of the lungs. The danger is that it is caused in a great majority of cases by the presence, or growth, of tubercle.

A question often arises in such cases, where does the blood come from. We distinguish this from the color. Blood from the lungs is arterial, and has a bright scarlet red; and is usually raised by hemming, hawking, or coughing; in short, it is *expectorated;* it is also, generally, frothy when in any quantity. Blood from the stomach, nose, throat, or tonsils is of a dark-red color, or venous blood. When from the stomach, it is usually vomited, or comes up with nausea, and is very dark, or quite black. When from the lungs it occasionally lies long enough to clot, or turn dark; but even then may be distinguished by a little cough, or tickling in the throat, instead of nausea and vomiting.

The signs which immediately precede bleeding at the lungs may be noticeable, or not. Generally there is a sensation of weight, heat, and oppression about the chest, and some difficulty of breathing; also signs of a congested state of the lungs, such as paleness and coldness of the skin and hands and feet, with chilliness or shivering. There is an unusual desire for fresh air just before the attack; and some persons subject to such attacks have warning, by being obliged to sleep with an open window, otherwise their breathing would become oppressed, and sleep disturbed.

The quantity of blood lost is very variable in these attacks. Sometimes the expectoration is merely tinged with it; and in such cases a doubt sometimes arises whether it does not come from the nose, mouth, or throat.   A considera-tion of the signs already stated will generally be sufficient to determine.   More commonly, in a first attack of bleeding at the lungs, the blood is coughed up, or hemmed up in considerable quan-tities, varying from a few drops at each expira-tion, to copious gushes by the mouthful.   And in such cases it is popularly, but wrongly, believed that "a blood vessel has burst."   *The quantity of bleeding is generally overstated, as bright-red blood makes a great show.*

The effects on the system of pulmonary hæmor-rhage are often trifling, and it even brings relief in some cases, being a natural effort to throw off oppression and embarrassment of the lungs.   In other cases it may produce great prostration; or, in the last stage, hasten death.

The treatment of pulmonary hæmorrhage will be considered in connection with that of pulmonary consumption.

------◆------

THE SYMPTOMS OF CONSUMPTION.

The progress of consumption is dependent upon the progress of the tubercular deposit in the lungs; but every educated physician, in comprehending these symptoms, will not fail to take into considera-tion the *physical signs*, or that information which may be gained, in regard to the state of the lungs, by placing the well-taught ear, or the stethoscope, against the walls of the chest.   This is popularly called *sounding the chest, or lungs;* and we, in this manner, connect the external and observable signs with those changes and alterations which morbid anatomy proves to be going on in the substance of the lungs. But since *sounding the lungs* can only be done by an educated physician, we shall omit in this work any consideration of this part of the subject, as foreign to a popular treatise.

The symptoms of consumption are usually considered under three stages, corresponding with three periods of tubercular formation. The first stage corresponds with tubercles in their crude, or formative state; the second stage, with that of their softening, or ripening; and the third stage corresponds with the period when they have softened, are coughed up, and large cavities have formed in the lungs. A small cavity so formed, and which might be curable, would not imply the third stage as present.

Symptoms of the First Stage.

The existence of consumption, at this period, owing to the obscurity of its symptoms, cannot always be detected with certainty. In fact the patient himself may feel so little uneasiness or anxiety, that he may be unconscious of any great departure from his ordinary health until the disease is far advanced, and the case has become desperate. In other instances, the symptoms are so prominent and so characteristic as to attract the attention of the most careless observer.

The symptoms and signs are materially modified by the age, strength, habits, and peculiarities of the individual: some may be altogether absent, others may be irregular, and all may vary in the degree of intensity. Although the symptoms in the first stage are usually obscure, and it is difficult to detect the real nature of the disease, we should always suspect the presence of consumption when we know there is hereditary predisposition; when we find a cough continue for some length of time, inducing increasing debility and emaciation; and especially when the invalid bears the appearance of a scrofulous constitution.

The commencement of consumption is slow and insidious; there is seldom any pain, in the part most affected, to direct the attention of the patient to his malady. After some slight exposure to cold, or other exciting cause, he feels an uneasiness at the back part of the throat, which induces a hard and dry cough; without being very troublesome the cough continues, and is soon accompanied by a trifling expectoration of frothy mucus, without color and without consistence, as in common catarrh. Presently the cough becomes more frequent and more decided, particularly in the morning on getting up, and at night soon after retiring to bed. The expectoration is now transparent, but more tenacious, almost ropy; any little exertion during the day, as walking fast, or going up-stairs, is sufficient to bring on a fit of coughing, and with it quickness of breathing, attended with some degree of oppression at the chest. The patient soon becomes sensible of unusual languor; he is readily fatigued, and finds his strength unequal to his customary labor or exercise; he breathes with some difficulty, and his respirations are shorter and quicker than usual; if he take a deep inspiration he is concious of uneasiness, scarcely a pain, immediately beneath the collar-bone, and this, more frequently, is felt on the right side.

The local disease now begins to implicate the general health; and, as the pulmonary symptoms advance, which they now do more rapidly than heretofore, the whole frame sympathizes with the chest affection.   The pulse becomes quicker than natural, especially towards evening; the body is frequently chilled with a sudden rigor, or shivering, which is followed by increased heat of the skin, particularly at the palms of the hands and the soles of the feet, which, towards night, are hot, harsh, and dry.   After midnight the feverish heat is succeeded by a moisture ; and, towards morning, the body is bathed in a profuse perspiration ; the sleep is occasionally disturbed by a sharp attack of coughing, and the patient arises, in the morning, relaxed and enfeebled.

The appearance of the invalid soon attracts the attention of his friends ; the countenance loses its healthy, rosy bloom, and at one time is pale and anxious, and again suddenly flushed with a blush of red ; the eyes sparkle with unusual brilliancy ; the hair grows long and damp ; the body diminishes in bulk, and begins gradually to waste ; the flesh loses its natural firmness, and is soft and loose ; the spirits are dejected ; the appetite precarious, and he is indolent, languid, and easily fatigued.

The patient may continue for a considerable length of time in the state just described ; he may gain renewed strength to combat the exhausting effects of his disease ; the further development of tubercles may be retarded by judicious remedial measures ; the growth of this, the first crop, may be arrested, and he may be restored to such a share of health as to remove the alarm of his connections. But too frequently the symptoms again return ; again they may be subdued ; and thus, battling with disease, life may be prolonged for years after the known and certain existence of that which at one time or other may prove fatal.   Dr. Rush relates that he knew one patient in this state twelve, and another twenty years. More cases, however, are now radically cured than in his day.

Frequently early in the disease, almost always towards the close, the tubercular affection involve the larynx, or organ of voice. The voice is more or less affected, sometimes entirely lost. When i occurs early, before the symptoms of affection o the lungs are distinctly declared, it often occasion: a delusive hope that this is all, and the attention i: entirely turned in this direction. A distinction here is important. If the symptoms are owing to a sim ple inflammatory affection, although they may be obstinate in their resistance, they are eventually very sure to yield to treatment. If they are owing to tubercle, they are very likely to go on to a fata termination. These two classes of the affectior usually go with the public, improperly, under the name of bronchitis, and are often uselessly very harshly treated, when a careful and intelligent ob servation of the symptoms would show that the af fection of the larynx is only one feature of a much more grave disease elsewhere.

It has already been mentioned that *pulmonary hæmorrhage* may be, and often is, one of the earliest symptoms of consumption. Such cases are among the least alarming, and, under scientific treatment most likely to result in a restoration to health, more or less complete.

It may be also remarked that *cough* is generally though not always, one of the earliest indications of incipient consumption ; and is consequently the first symptom which excites the attention of the patient or his relations. During the first weeks, or months and in some cases years, it may be slight, occurring mainly in the morning. It appears often to arise simply from irritations in the region of the throat hence it does not excite alarm. We would by no means frighten people at every little symptom of cough they may experience. I would not have them unreasonably alarmed and imagine that, because they have a cough, they must necessarily pass into consumption ; but, at the same time, I would have every one remember that *coughing is never a symptom of health.*

A person who coughs is not entirely well ; and consequently no pains should be spared to arrest, if possible, the symptom ; and in no one thing is it more important to begin in season. As Poor Richard says, " A stitch in time saves nine," and if this principle is an important one in the common affairs of life, how much more so in matters of health.

Symptoms of the Second Stage.

These cannot now be in doubt. The cough, which before was only occasional, is now frequent and distressing. The expectoration is no longer a clear, scanty, frothy mucus, but is more abundant, and assumes the character of pus, or matter ; which presents, on examination at different periods, part, or all of the following appearances:

It is thick, opaque, and of a pale-yellow color ; sometimes it has a greenish tint, and at others it is dark, almost black ; a portion may acquire a greater, even hard consistence, and be surrounded by a watery or whey-like mucus ; it may be tinged with blood, or contain small specs or streaks of blood ; small solid particles, or shreds, resembling curd, of a dead white or straw color, varying in size from a pin's head to a grain of rice, may be noticed floating or sustained, either in a cream-like or a transparent fluid ; sometimes the softened tubercles are coughed up in flakes. The expectoration, in some cases, is devoid of smell ; in others, it

has a faint fœtid odor; it is of greater specific
gravity than water, and, when deposited in a ves
sel containing that fluid, mixes with it, or sinks to
the bottom.

The cough, although constantly tormenting the
patient, is seldom attended with any acute pain
except when there is some slight degree of inflam
mation of the *pleura* (the investing membrane of
the lungs, and the lining membrane of the chest)
or when old adhesions of the two *pleuræ* — the re
sult of former inflammation — interfere with the
natural expansion of the lungs. Pain, almost of
rheumatic character, — indeed, it is sometimes re
ferred to rheumatism alone, — is frequently experienced around the shoulders, between the shoulder
blades, and at one or both sides; occasionally, there
is difficulty in lying in bed on one or the other
side, without some pain and uneasiness. In general, the amount of pain endured, during the progress of the disease, bears no proportion to the extent of mischief going on in the lungs.

The difficulty of breathing, which in the first
stage was temporary, is now, in the majority of
cases, constant. This may be readily accounted
for by the increased size and increasing number
of the tubercles having encroached upon, and
blocked up the air-cells, and thus diminished that
surface of the lungs by which the act of breathing
is performed.

Hectic Fever.

When the expectoration is purulent, and presents the characters I have just described, that
condition of the system which is designated hectic
fever, always prevails; at the very commencement
of consumption, this fever slowly and insidiously
affects the health and strength; but it is seldom

that it manifests itself, in all its fearful symptoms, until the tubercles begin to liquify and pus is formed.

Hectic fever is of a remittent type, and is said to have two accessions in the twenty-four hours: one in the middle of the day, and the other towards evening; with the exception of the evening increase, which is always regular, the periodicity of its return is uncertain; sometimes it is absent altogether during the day, and sometimes the patient is never free for any length of time from its sudden invasion; but these repeated attacks are never so severe as that which exhausts the patient in the evening and night.

The access of the fever commences with chills and shuddering, and a sense of "creeping" in different parts of the body; after a time, varying from half an hour to two or three hours, the *hot stage* succeeds, and the patient is then burnt up with fever; he is restless, and overpowered with lassitude; the pulse is seldom less than one hundred — more frequently one hundred and twenty; the skin is hot and dry, and the face is flushed and burning. This stage lasts several hours, and towards morning terminates in perspiration.

The ordinary acceptation of the word "perspiration" is quite inadequate to express the amount of the *night sweats;* the body is not bedewed, or damp, but wet; perspiration, like drops of water, oozes from the pores of the skin, and, in some instances, rolls from the body almost in a stream, so that towards morning, the personal clothing and bed-linen are completely saturated with moisture. Of all the signs diagnostic of consumption, not one is so constant, or so confirmatory of the disease, as these night sweats.

When hectic fever is established, the pulse increases in rapidity, and beats from one hundred to one hundred and twenty or one hundred and thirty; the heart palpitates violently, and is easily excited by trifling causes; the respiration is hurried; the cough is "hacking" and exhausting; the body loses flesh, and wastes or *melts* away; the flesh that remains is soft and flabby, and the skin loses every appearance of health. The debility is great, and the lassitude so increases that the patient is quite unequal to any bodily exertion. The sleep is invariably disturbed by repeated paroxysms of cough, induced by the loaded state of the air-passages; and the least change of position, as turning from one side to the other, is sufficient to cause a recurrence of the attack. The appetite is fickle; sometimes it remains good to the last, but more frequently there is perfect loathing of food, which occasionally produces nausea and vomiting; thirst is seldom troublesome or excessive, even during the feverish state. At the commencement, the bowels are usually constipated; after a time they become irregular, being relaxed for several days, and again costive; when, as may happen towards the close of the disease, the mucous membrane of the bowels is irritated, or even ulcerated, diarrhœa is frequently present, and greatly assists in reducing still lower the remaining strength of the sufferer. The urine is generally variable in quantity, and deposits a bran-like sediment.

Pulmonary Hæmorrhage frequently becomes an alarming symptom at this stage of the disease, if it has not before occurred; and by presenting to the patient visible evidence of the existence of internal mischief, excites the first suspicion in his mind of the reality of his case.

Menstruation is either irregular, deficient, or quite absent; and this suppression is often wrongly considered as the *cause* instead of the *effect* of the pulmonary disease. This suppression may also occur in the first stage.

The general appearance of a sufferer at this stage of decline is so typical of the disease, that, to the experienced physician, the face and figure depict, almost describe in detail, every symptom.

Dr. Rush gives this description : —

" The nose becomes thin, especially at this point ; the check-bones project ; the skin covering them is pale during the day, in the evening it is flushed in circumscribed patches of a brilliant red color (hectic blush) ; the white part of the eye shines, and is of a light pearly hue ; the eyes are large and bright, although somewhat sunk in their orbits ; the cheeks are hollowed ; the lips retracted, presenting often the appearance of a melancholy smile ; the the teeth increase in transparency ; the whole body is shrivelled ; the spine projects, instead of sinking, from the decay of the muscles ; the shoulder-blades stand out like the wings of a bird ; the fingers are shrunk, except at the joints, which are prominent ; the nails are curved ; and the hair gradually falls from the head."

During this wreck of health the mental faculties continue perfect, and are often endowed with increased intelligence ; the temper may be occasionally irritable, but the spirits are seldom oppressed on account of the malady. Hope, a strong hope of ultimate recovery, constantly and wonderfully sustains the patient ; he will admit he has " a cough which may be serious ;" but " when warm weather comes he will be better." How often have we heard a girl, who could scarcely utter the words ; " Wonder why mamma was fretting ? " unconscious that her own danger was the cause of a mother's sorrow.

Symptoms of the Third Stage.

This stage of consumption coincides with the complete softening of the tubercles, when their now fluid substance finds its way into the bronchial tubes, and is gradually expectorated, and a cavity occupies the place.

The symptoms described as characteristic of the second stage, now prevail in greater intensity; the cough is scarcely absent for any length of time, but tears and racks the breast, sides, and back with sharp, lancinating pains, and leaves the patient, after each paroxysm, faint and exhausted; during the night the cough is unceasing, and drives off that natural and blessed restorative — sleep. At the commencement of a paroxysm, the cough is "hollow," but as the expectoration becomes loosened it gives a gurgling, or rolling sound, which gently subsides almost to a murmur. The expectoration is profuse, occasionally amounting to a pint in a few hours; it consists of a heavy, purulent discharge, in consistence equal to cream, and in color varying from pale yellow to green, or bluish-black or brown; it contains small lumps of a curd-like substance, and is sometimes freely mixed with fresh florid blood; at others, the blood is in minute congealed clots or threads; the odor is generally faint and sickly, in some cases foetid and offensive. The expectoration may be so copious in quantity, and the strength of the patient so prostrated, as to deprive him of ability to eject or cough up the accumulated matters, and thus suffocation may be threatened. A case occurred at Paris, during the time the editor was there, in which death was instantaneous from these causes.

The breathing is oppressed and difficult. The difficult breathing does not come on in occasional or spasmodic attacks. but is constantly laborious, in

consequence of the imperfect inflation of the lungs
—perhaps I should say, of what remains of the
lungs; the least exertion, or change of position, aggravates the oppression, and the sufferer obtains
breath by a succession of gasps, rather than by natural respiration.

The hectic fever ravages the frame with undi-
minished violence; the chills are frequent; the succeeding heat produces exhausting faintness, and
the perspirations during the day, as well as the
night sweats, are abundant. Diarrhœa is generally
present, and the copious evacuations which are
constantly occurring, reduce the strength of the
patient to the lowest possible ebb, and constantly
cause an overpowering sensation of faintness and
sinking. The appetite is bad; and it is only by
the most savory, delicate, and not always the most
proper food, that the patient can be tempted to eat.
Whatever is eaten readily causes uneasiness and
disturbance in the stomach; sometimes it is quickly
rejected; but, if retained, it creates so much irritation as to produce pain and nausea. Flatulence,
and violent eructations of acid wind, constantly
harass the patient, and occasion a "rising in the
throat," which appears to threaten suffocation.
The pulse retains its unnatural rapidity, and is seldom less than one hundred and ten; the surface
of the body is always hot to the touch, and the
palms of the hands and the soles of the feet are
burning. The throat and mouth are generally sore
from numerous small aphthous ulcers, and, in some
cases, the larynx is ulcerated: when this occurs, it
renders the cough still more frequent and painfully
distressing. We have, in several instances, noticed
the formation of small abscesses, either in the rec-
tum, or in the immediate neighborhood of the lower bowel, during the last stage of consumption; in-

deed, the whole mucous membranes appear to approach closely to ulceration, if they are not absolutely ulcerated.

Towards evening the feet and ankles become swollen, tumid, and filled with fluid, and dropsy in various forms may make its appearance; sometimes the limbs are dropsical, at others the abdomen is tumid, or the chest fluctuating. When dropsy becomes general, as is sometimes the case, the night sweats and diarrhœa cease; within a few days, however, the perspiration may return, and then the infiltration subsides, so that one set of symptoms alternates with the other.

With these symptoms the emaciation and debility keep pace; the strength is barely sufficient to support the limbs, and the frame is reduced to the condition of a skeleton; the joints are large and protuberant; the nails grow rapidly, and become more incurvated, almost like talons; the hair is damp, weak, and continually falling. The voice, when the larynx is ulcerated, is hoarse, and attended with a clanging sound; sometimes it is shrill and hollow, and at others the patient can scarcely speak louder than a whisper.

Whilst the physical powers of life are thus decaying, the mind holds its preëminence unimpaired; the faculties are acute; and, strange as it may appear, are capable of the highest cultivation, and even of abstruse study.

To the last moment the sufferer still clings to hope; is unconscious of any inward emotion that tells the disease is fatal; views the despondency of friends with surprise, almost with peevishness, and is ever buoyed up with the faith, almost the certainty, of recovery.

In other cases, but they prove the exception rather than the rule, the mind is comparatively torpid;

the patient is indifferent to a return of health, or to a fatal issue; and in some cases an excited delirium attends the last days of life.

Although the course of the last stage of consumption is characterized, in a large majority of cases, by the symptoms now detailed; yet, in some instances, there may be a total cessation of those prominent and peculiar signs which belong alone to the close of this devastating disease; thus, we may occasionally see cases in which the cough, the expectoration, the diarrhœa, the exhausting perspirations, cease altogether, and leave the patient in a state of happy and placid tranquillity. When this occurs, it must be attributed more to the failure of the animal powers, and deficiency of material, than to any permanent restoration of the system; for at last the scene closes by life gently gliding away, "like the expiring lamp," in ease and peace.

TREATMENT OF CONSUMPTION.

The opinion has already been expressed that consumption is a disease, whose essential nature is *scrofulous*, or *tuberculous*, and commences by the formation of morbid matter in the blood. But it is also a disease of debility; of imperfect nutrition; and usually attended by excessive irritability of the nervous system.

The especial aim, therefore, of all treatment must be to purify and eradicate this morbid matter from the blood; to promote the absorption, or elimination of tubercles when already formed; to assist the healing, and cicatrization of cavities; to prevent the further deposit of tubercles; and to support the strength and improve the general state of health, while all this is being done. In short the *attempt* should be made to *cure* the disease; and if many cases are incurable, that is no reason why the consumptive should not take his chances for a cure under new and improved modes of treatment; more especially when the same means will result at least in temporary comfort and prolongation of life, if not in a permanent cure.

The first indication, then, is to reinvigorate the debilitated system by purifying, restoring, and enriching the blood. This cannot be done without a plenty of *pure air*, and simple, but substantial food. But how can food be digested, and properly converted into blood when the digestive organs are debilitated, and the blood and whole system contaminated by the scrofulous or tuberculous habit? We answer, give a medicine which is adapted to this very purpose. It is called Rush's Sarsaparilla and Iron; and will be found decidedly the best remedy for this purpose, during the first and second stages of this disease. Take a teaspoonful *after* breakfast and tea, and two teaspoonfuls after dinner. It will result in a gradual, but sufficiently rapid improvement in the appetite, and blood-making powers of the system, and should be continued as long as it benefits. See page 47.

It is hardly necessary to say that the old-school remedies of *bleeding, cupping, and leeches, digitalis, Prussic acid;* and, in case of females, of forcing remedies to bring about the monthly periods, should be among the things of the past.

Bathing, in the first and second stages, and particularly while taking Rush's Sarsaparilla and Iron, will be found very serviceable. The morning sponge-bath, see page 13, and also the same bath taken on going to bed at night, if there be any cough, restlessness, or tendency to slight feverishness or sweats, will be the best and most convenient bath. In case of constipation, or of females with suppressed menstruation, the sitz-bath at 70 degrees beginning at 75, page 30, for fifteen minutes before going to bed, followed by rapid sponging of the rest of the body, before getting out of the tub, will often be productive of a good night's rest, and great comfort and relief.   The baths here recommended should not be abandoned as soon as temporary improvement sets in ; but should be continued even through cold weather, until, in connection with the remedies herein recommended, a cure if possible is effected.   The reader should not fail to consult the article on *Diet*, at page 15 ; the article on *Laws of Health*, page 8 ; *Water Treatment*, page 24 ; and *Nursing*, page 27.

The diet should be nourishing, without being stimulating, and should be strictly confined to the kinds mentioned under *Diet*.   The following is a sketch of a day's diet on a consistent plan : —

Breakfast. A large cup of chocolate (unspiced is the best), or weak, black tea, or half a pint of new milk, with Graham bread, or bread made of wheat meal unbolted, and toasted if preferred, a little lightly salted butter, and a soft boiled egg, if eggs agree, or a basin of oat-meal porridge and cream ; or, wheat is better, boiled wheat and cream.

Dinner. A broiled muttonchop, or beefsteak ; broiled or roasted chicken, or roast beef, or mutton, rather rare ; mashed potatoes and cream, and such allowable vegetables as agree ; bread as at breakfast ; wheat-meal pudding and cream, or boiled wheat pudding and cream, or instead of cream, if that can not be had, maple or Stuart's syrup. A cup of weak black tea is allowable, but not advised. Pure water is better. Raw or baked apples are allowed.

Tea. A weak cup of black tea, with Graham bread and butter, or boiled wheat and cream, or boiled rice and cream, or sweet baked apples, or potatoes and cream. Simple preserved fruit is allowable. Eat it with cream.

If the bowels are constipated, the boiled wheat and Graham bread, or wheat-meal puddings are excellent, and are highly nutritious, and at the same time unstimulating. Sweet cream is an excellent article for consumptives, and when it can be obtained, should be taken, in moderation, at every meal. It is to take the place of that odious "cod-liver oil," once so much used. If at any time there should be diarrhœa, rice and flour bread may be substituted for boiled wheat and Graham bread. A glass of pure cold water, on going to bed, is an excellent remedy for restlessness or tendency to constipation or fever and sweats. A teaspoonful of Rush's Sarsaparilla and Iron may be taken at the same time, for the same purpose.

Consumptives cannot always bear as ample a diet as the above, at first: but there will be but few cases which will not come up to the above standard, after taking Rush's Sarsaparilla and Iron and Rush's Lung Balm, as hereafter recommended, one or both, during two or three months.

If *pure air, bathing,* and *diet* are important, proper exercise is hardly less so. *The very best* is riding on horseback; but walking, riding in a carriage, or sailing will do. It should be taken regularly, and not lightly given up on account of a raw or cold day. A moderate course of gymnastic exercises will often assist to give energy and vigor to the system. They should be begun moderately, and increased by degrees, so as to call into action all the muscles of the body in turn. When a regular gymnasium is accessible, it should be daily frequented; but such motions as require *raising the arms above the head,* should be avoided, especially in cases of bleeding from the lungs. And this caution applies to daily household matters. When the strength of the patient will not admit of any great exertion, *swinging* in the open air, or in bad weather within doors, is a healthful and soothing recreation.

Graces and Shuttlecock are exercises that call into play the muscles of the chest, trunk, and arms, more especially; and are therefore good for those predisposed to consumption. If females will persevere, and learn to play with the left hand, as well as with the right, they will find it an excellent mode of preventing spinal curvature, and thus of invigorating not only the whole spinal column, but the system generally. They should be practised in the open air.

We might give this rule:—Exercise as much as may be without much fatigue. By this means, the system becomes invigorated, and warmed; and as more perspiration results, more *pure water as a drink* will be required. This will be an advantage, and *water drinking should be encouraged;* six half pint tumblers a day, besides at meal times, is not too much; but many persons must begin more gradually, and increase in the same way. Exercise, then, and drink as much water as you conveniently can.

Spinning wool at the old-fashioned hand-wheel, used to be recommended as a very useful exercise. Perhaps the more modern sewing-machine will now *do*, if light and easy running; but anything like a day's work is too much for most consumptives, at least until they are strengthened by purification and renovation of the blood, by suitable medicines, and a correspondingly substantial diet.

Light work on a farm, or garden, avoiding lifting and violent exertions of strength, will be found highly beneficial; and out-of-door pursuits always preferable to those within doors; and useful forms of exercise, to those which are taken merely for the sake of exercise.

The subject of *clothing* has already been briefly noticed. Clothing should be warm, and sufficient, without being relaxing; and so regulated as to preserve the surface of the body at an equable temperature. The question of wearing flannel next the skin, is one which has been much discussed. We do not recommend it for consumptives, or others. It absorbs the perspiration, becomes damp, and does not readily part with the moisture by evaporation. It retains all the morbid secretions constantly exuding from the body when diseased; and, in many instances, creates that degree of heat which is too relaxing, and always tends to diminish that hardihood of constitution which is the best preventive of disease. Consumptives, however, who have long worn flannel, cannot throw it off at once, especially during the cold season; but would do well to put it on *over* the cotton, or linen shirt, which should then be placed next the skin. The consumptive should never sleep in the same clothing he has worn during the day.

TREATMENT OF COUGHS.

Cough is one of the most common and troublesome symptoms not only in consumption, but in several other complaints. Preparations containing *opium*, or its active principle *morphine*, are the most in use, not only among old-school doctors, but also as forming the base of nearly all the advertised remedies for coughs, colds, and consumption.

Ayer's Cherry Pectoral for instance, we believe, contains *morphine;* and we might almost go over the whole list, without finding an important exception. Dr. Rush says, in regard to this: "I would earnestly express my disapprobation of the too common practice of obtaining temporary relief from *opium*, or *morphine*. This drug quickly loses its power of doing good in innocent doses; so that the quantity necessary to produce the desired effect must be daily augmented, until it becomes no longer innocent. I may add, that opium is the basis of all the quack advertised nostrums for cough, asthma, and consumption; the increasing supply which the system demands when once habituated to its use, is not the least favorable point to such *mercenary speculators.*"

In the earliest stage of the disease, the cough is seldom very troublesome; and it is in this stage that sufferers are earnestly advised to beware of preparations of opium. Very few, *if any*, who have tuberculous taint, and begin with such preparations, are ever able to discontinue their use; and they *prevent* the effectual use of any really curative remedy. It will be far better to seek temporary relief from bland, demulcent remedies, such as the refined extract of liquorice, linseed, or slippery-elm tea; or what is better, hoarhound, or boneset candy; the candy, however, to be used in moderation, only just enough to produce the desired relief. But in such a disease as incipient, or threatened consumption, — and *every* cough threatens consumption in consumptive families, and in those persons who have the signs already described as indicating scrofula, — the *only* safe way is at once to rid the system of this scrofulous, or tuberculous taint, by the use of appropriate remedies. The *best* and *only* remedy, yet discovered, which will unfailingly cure coughs, and at the same time promote the removal of tubercle and the tuberculous diathesis, and scrofulous habit of body from the entire system, is Rush's Lung Balm. It should be taken in doses of half a teaspoonful, or thirty-five drops, every six hours; one of which doses should be at bedtime.

If the patient is taking Rush's Sarsaparilla and Iron at the same time, as previously directed, that should also be continued, and never interferes with any other remedies. The advantage of taking the two medicines together is, that the Sarsaparilla and Iron is more energetic in purifying and renewing the blood; while the Lung Balm is most effectual for the cough, and in promoting the removal of tubercle. For an account of the discovery and perfection of this remedy, see page 54.

Cough, in the more advanced stages of consumption, needs still more the assistance of this remedy, which consumptives will find, on trial, to afford a far more grateful and permanent relief than any other medicine.

The expectoration, when at all abundant or thick, heavy, greenish or yellowish in color, as described on page 111, will be benefited by the addition of 60 grains of the hypophosphite of lime, or the hypophosphite of soda, to each bottle of the Lung Balm. It can be procured at any druggist's, or by writing to the editor (for 12 cents). Dissolve it in half a teacupful of boiling water, and mix it with the bottleful of Lung Balm. It may be mixed in a pitcher, and stirred up, and immediately poured back. The remainder, which the bottle will not hold, may be put in a smaller bottle and used first. This addition to the Lung Balm is not essential, but only advisable in cases where there is the above kind of expectoration. The Balm may be taken without it, with great benefit in all cases. The dose, after the above medicine is mixed, will be forty-five drops. Rush's Lung Balm contains no opium, or morphine, or any preparation of them.

One of the best palliative means for cough is to take a sponge-bath with pure water at 70 degrees, over the whole surface of the body, and more especially the chest and throat. This will be found a cooling relief in cases where there is slight feverishness, and the skin hot and dry. Even bathing the feet will often relieve a troublesome cough. The sipping frequently of pure cold soft water will frequently assist in removing a troublesome cough.

The difficulty of breathing which attends consumption may be greatly modified, and if the case has not gone too far, quite cured, by Rush's Lung Balm, as before directed. It may be also temporarily modified and relieved by bathing the chest with cool water, and abundant rubbing with the wet hand, two or three times a day, or oftener if agreeable.

The treatment of *Pulmonary Hæmorrhage*, or *bleeding at the lungs*, has been very much debated among physicians. It used to be the fashion to bleed. But that is now mostly abandoned, even by old-school men.

Dr. Rush had a favorite remedy, which he made use of to check the bleeding, and which may now, in this country, be considered a popular remedy, namely, common salt, taken in a dose of a teaspoonful every few hours, till the hæmorrhage abates. This article, although so common, is a powerful drug when taken in such quantities. I should be unwilling to take it myself; but it may answer a temporary purpose, with the view of arresting the flow of blood, when no other remedy is at hand.

But the best remedy is Rush's Lung Balm. Put a table-spoonful in a tumblerful of water, with some pieces of ice, if it can be had, and give a tablespoonful every half hour until the hæmorrhage is arrested, or for three or four doses; then once in three or four hours, and after the first day in the usual dose of 35 drops every six hours.

The feet, during an attack of this bleeding, will usually be found cold, and should be put in very warm water, reaching above the ankles, until they are thoroughly warmed, and then be kept so by hot bricks. At the same time make cold applications all round the front and sides of the chest, by means of napkins wet in very cold water, kept cold by ice, if it can be had, and changed very frequently, with the view of cooling the whole of the lungs and the mass of the blood. There will be no danger of taking cold from the abundant use of cold water so applied; particularly *if the feet are kept warm.* There is no doubt of the great utility of these cold applications to arrest such bleeding; even pounded ice in folded towels may be used in some cases. The Lung Balm should be continued at least three months after an attack of bleeding at the lungs. And if there is decided debility, Rush's Sarsaparilla and Iron should be taken at the same time, a teaspoonful after each meal; the Balm 35 drops every six hours, avoiding meal times. One bottle of the Sarsaparilla and Iron will generally be sufficient.

Hectic Fever. The appropriate use of water in all fevers is said to be the luxury of the rich; but the poor may also use it, if they will but throw aside their prejudices and learn how. As soon as the sufferer begins to feel the heat coming into the hands and feet, he should take a sponge-bath all over; see page 13; or if that is inconvenient, let him wash freely the face, hands, arms, and feet, and he will not fail to be paid for his time and trouble. Heat will be prevented, and the strength supported by this plan ; and debilitating sweats probably avoided. The *cooling treatment*, too, has a great influence in regulating the circulation, and preventing rapid pulse.

Exercise, which we have spoken of as so valuable, should be so regulated, if possible, as not to hurry and excite the pulse. And here it is worthy of remark that *over heated rooms* always excite the circulation, and make *hectic fever* worse. A thermometer is indispensable in the sick-room, both to determine the temperature of baths, and that of the patient's room, which should be kept at from 65 to 70 degrees, with one window at least, let down at the top, during the cold season, for the sake of pure air; with open windows in summer; do not be afraid of that bugbear, " taking cold." How often have we found consumptives literally sweltering to death, with hot stoves, and the air close and foul with the morbid exhalations from their bodies. Consumptives will not recover, under *any* course of treatment, shut up night and day in an unventilated room. They say, " If we open a window we shall take cold." We say, " No you will not, if you open your window before you make your fire, and keep it open all the time; or if, in other words, you will keep your room at a uniform temperature ; " but this of course will require some expense for fuel, and some care. But remember, consumptive reader, that you will die under the old way, while you *may* get well if you do as here directed. Keep your thermometer where you can see it all the time, and don't let it get any over 70 degrees ; and keep away from the stove.

The *night sweats*, which are often excessive and very prostrating, may generally be controlled by taking a sponge-bath three or four times a day, as previously directed, when the skin is hot and dry.

9

One may also be taken on going to bed at night, and if that is not sufficient to check it, take *another* as soon as the sweating begins, which last will rarely if ever fail.   Rush's Lung Balm is, however, the sheet-anchor in all these states; and all other treatment is only as an aid.   For the night sweats take a dose on going to bed, and another at the time the perspiration comes on; also twice during the day, say at 10 A. M. and 4 P. M.   The Hypophosphite of Soda should be added as directed previously for cough.   The Sarsaparilla and Iron should be taken after each meal.

Pains in the chest may generally be relieved by applying a wet towel wrung tolerably dry from cool water, putting this folded over the seat of the pain, keeping it wet and covered with dry flannel enough to keep pleasantly warm.   In hot weather very little covering will be required, and more frequent wetting.   A small napkin can be used first, and afterwards a larger one : begin moderately, and gradually increase.

The diarrhœa of consumption is always formidable.   It may be generally very much relieved by injections of cool water into the bowels after each motion.   At least a pint or two of water should be thrown up with a pump, and retained some minutes, or as long as the patient can bear.   Take the Lung Balm every six hours, as previously directed.

Leucorrhœa, or *Whites,* is a very common trouble in females, during consumption   It should be treated as directed under that article.

In regard to the use of baths, and water for bathing, it may here be added that the addition of a little of Rush's Restorer to each basin of water used. say as much as can be held on the blade of a penknife, will prevent taking cold, and relieve a feeling of soreness and fatigue.

Treatment of Consumption by Inhalation.

When, and to what extent is inhalation available?
If *extensive* cavities have already formed in the pulmonary tissue, we cannot in the majority of cases, look for any permanent improvement, whatever may be the means employed. Still, in those unfavorable cases, surprising relief will often be afforded by inhalation in connection with other suitable treatment, and life may in many instances be thus materially prolonged.

Such persons should begin the practice with caution, since whatever tends to over-exercise the lungs, under such circumstances, will only tend to hasten on more rapidly the fatal work. Any such danger may be avoided by following suitable directions, and the use of a proper inhaler.

Inhalation performed two or three times daily, and half an hour or an hour at each time, has been found in a few weeks to work a wonderful change for the better in the chest. "Externally, the muscles concerned in respiration become manifestly enlarged," says Dr. Rush, "and the bony walls of the chest, both before and laterally, visibly increased; while at the same time the natural respiratory murmur will be heard internally far more distinct than ever. Such has been the increase of size which the chest, in young persons especially, has undergone by practising inhalation, that I have known individuals, after inhaling little more than a month, require their waistcoats to be let out. It is a fact incredible to one who has never been at the pains to measure the chest, or examine its shape," this author also says, "what an enlargement it acquires by the simple action of breathing, for the time above stated (a half hour, two or three times daily), backward and forward through a narrow tube of a few feet in length."

There has, however, been an immensity of deception practised in regard to inhaling vapors and gases, as remedial agents. The main benefit results to the lungs from exercising and expanding them, and affording an increased supply of *pure air*. Enough, however, has already been said on the subject of the capital importance to consumptives of a supply of *pure air, without which* no remedies can ever purify or renovate the morbid condition of the blood.

The vapor of the same medicine that the consumptive is taking can always be inhaled to advantage by means of a suitable inhaler, of simple but effective construction. Rush's Lung Balm and Rush's Restorer are the medicines best adapted for this purpose, and have, in several cases, contributed to highly satisfactory results. Patients are much more likely to use an inhaler perseveringly if combined with a medicine which has a soothing and grateful feeling to the sore and debilitated organs. Such is the effect of the remedies above mentioned; and we venture to say that no person who has once tried them, with a suitable inhaling apparatus, would be without the relief and benefit enjoyed for the expense and trouble they cost.

Suitable inhalers unfortunately cannot be obtained at most drug stores, and the editor of this book has been obliged to keep on hand a supply, which he will continue to furnish at as low prices as the market will admit of. Full directions for their use accompany the inhalers. For the editor's address, see page 44.

Consumption Curable.

By this is not meant that *every* case is curable, but that *some* cases are so; and it would be safe to say *many* cases if they could be early treated, and under favorable circumstances. The instances of cures in this book are given not in a boastful spirit, but for the sake of encouraging the faint-hearted to persevere in a rational and comfort-giving mode of treatment; and one which the editor candidly believes presents the *best*, if not only chance of success in this most formidable disease. For cases of cures, see page 58.

Persons desirous of corresponding with the editor of this book on any medical subject will find his address at page 44. Advice is cheerfully given, in all cases, free of charge.

PREVENTION OF CONSUMPTION.

In considering this very important subject, I shall commence at the origin of the evil, and this, in an immense majority of cases, is *Hereditary Transmission.*

It would be foreign to this work to discuss the hitherto inexplicable power which man possesses, of transmitting peculiarity of talent, of form, of defect, in a long line of hereditary descent; we must be contented with the *fact* that he has that power — that wit, beauty, and genius, dulness, madness, and deformity, are thus propagated to a future lineage; and that a host of fearful diseases, as gout, consumption, scrofula, and leprosy, originating perhaps in the first sufferer accidentally, are propagated so deeply and so extensively, that it is difficult to meet with a family whose blood is totally free from all hereditary taint.

The health of the parents influences the health of the child. Dr. Rush says, "The importance, therefore, of considering the health of the parent as the most effectual means of checking the extension of consumption, must be admitted; and I fear we must be content with the admission. **Is a thought** ever bestowed on this subject in **matrimonial alliances ?**"

Our opinion is, that when *both* the man and woman are tainted with a tuberculous constitution, marriage, under such circumstances, should be forbidden by prudence, if not by civil rule.

When a disposition to consumption exists in a family, "There can be no question," says Dr. Rush, "that intermarriages among the collateral branches tend more than anything else to fix and multiply and aggravate it; there is reason to believe that unions between total strangers, and perhaps inhabitants of different countries, form the surest antidote."

He also advances the opinion that the physical and moral constitution of the infant has a greater resemblance to that of the father than that of the mother. If this be correct, the health of the infant would be dependent in a greater degree upon the health of the father than the mother. The doctrine, however, in relation to form, complexion, and moral character has so many exceptions, that its correctness seems doubtful. Be this as it may, the young mother should know that the health of her infant depends much on her own; and that, from the commencement of pregnancy, she must consider herself responsible, to a great degree, for the health of her offspring; whatever interferes with the regular action of her several functions, especially digestion and its product nutrition, interferes with the growth, the development, and the constitution of the child yet unborn; and irregularity or carelessness, at this period, may entail upon her infant the most dire afflictions.

We will now consider the prevention of consumption in infancy and childhood, and the means by which we may improve the constitution, so as to overcome the hereditary predisposition. Our help-mates, whilst the infant is "mewling and puking in the nurse's arms," are proper diet, pure air, and bathing. If the child derive its consumptive con-stitution from both parents, or from the mother only, the latter must be deprived of her sweetest privilege — that of suckling her own child; if, on the other hand, the predisposition be acquired from the father, and the mother's health be unexception-able, this restraint need not be imposed. Food of "Nature's cooking, a mother's milk," is the natural sustenance of infancy. When a stranger's breast has to afford this, the greatest care is demanded in the selection of the "wet nurse:" she must be healthy herself, and of healthy parentage; in age she should not exceed thirty; her child should not be more than six or eight weeks old, and her tem-per should be good and placid, as the secretion of milk is naturally affected by irritability and pas-sion.

It is a common error with healthy mothers to suckle their children for twelve, eighteen, or twenty months, to the risk of their own health, and the injury of the child. Soon after the appearance of the teeth, the stomach of the infant is capable of digesting artificial food, and the milk of the moth-er is, after the eighth or ninth month, deteriorated in quality, and insufficiently nutritive; the child should then be weaned.

In consequence of ill-health, disease, or death of the mother, it may become compulsory to rear the children "by hand," — that is, entirely on prepared food; and certainly this mode, hazardous as it is, is preferable to nursing with the milk of a parent affected with consumption. An artificial milk, which approaches in quality that of the mother,

may be made with two-thirds of cow's milk and
one-third of water, to which a little sugar is to be
added; this forms a good substitute, and should be
made fresh as often as the child requires it.   The
French prefer diluting cow's milk with an equal
quantity of fresh whey.   Biscuit, powdered and
boiled with milk, water, and sugar, is also well
suited to the delicate stomachs of infants.   Arrow-
root, of all vegetables, is the least disposed to fer-
mentation, and forms an excellent food, either with
milk, or with water and sugar.   It is very common
in this country for people to give their children the
worst food possible, namely, flour boiled in milk,
which, when taken into the stomach, ferments, and
fills the intestinal canal with wind and acidity.
Not any animal food should ever be given to an
infant under nine months old.

Happily, the day has gone by when the new-born
babe was swathed and rolled in flannels and band-
ages, until deprived of all power of motion ; yet,
at the present time, dear old grandmammas adhere
too closely to the unhealthy custom of *their* childhood, and "long clothes," rollers, and nightcaps
still improperly maintain their place in the nursery.
It is a sadly mistaken notion to suppose that we can
give strength to a delicate and puny infant by
keeping it constantly in an artificial state ; an infant
confined in a heated chamber, lumbered with a
superabundance of clothing, must of necessity become so tender and susceptible, as to take cold upon
any and every alteration of temperature to which
it may be exposed.

In the early infancy of children, we must en-
deavor to adapt the feelings and constitution of the
child to the climate and circumstances by which it
is surrounded, rather than accommodate and regulate the atmosphere and dress to the supposed lim-

ited endurance of the child; our aim being to give to the infant an innate and native power of resistance, to render it a hardy perennial, not a tender hot-house annual. The clothing should be sufficient to preserve the body at a proper warmth, but not abundant or heavy; bleached cotton is the only fabric to be worn next the skin, and this should be changed frequently; and the child should be daily washed or plunged in cold water, and a genial reaction induced by gentle rubbing with towels. The importance of pure air has already been stated.

In boyhood, the diet should be nourishing and generous, without being stimulating; animal food should be given very moderately, *without pork*, and vegetables should be allowed abundantly. Exercise in the open air must be obtained at all seasons. Exercise at this age is a natural want, essential to train the muscles to their requisite offices, and to insure to the frame its full development and just proportions. So strong, indeed, is this tendency to motion, that few punishments are more grievous to childhood than such as impose restraints upon it. Girls should exercise out of doors as well as boys.

Little bodily restraint should be imposed on children for the first six or eight years; long and irksome confinement to the sitting, or, indeed, to any one position, and especially in close rooms, cannot but be inimical to the just and healthy development of their physical constitution. It is better that they be allowed to choose their own muscular actions — to run, jump, frolic, and use their limbs according to their own inclinations; or, in other words, as nature dictates — than to be subjected to any artificial system of exercise. In children of weakly constitutions, severe mental application is, in a particular measure, hazardous. Whenever a precocity of intellect, or a disposition to thinking and learning in advance of the years, is displayed, to the neglect of the usual and salutary habits of early life, it should be restrained rather than encouraged; the physical education should ever be of paramount regard; the future health — for the absence of which life has no recompense — being closely dependent on its judicious management. The practice, unfortunately too common, of selecting the most delicate child for the scholar, is founded in error. This is the very one whom it becomes most necessary to devote to some calling which demands physical action and exposure to the open air.

A proper and moderate use of the vocal organs, at this age, is of considerable advantage : reading aloud is the best method of training the voice and expanding the lungs; and if, at the same time, the pupil be taught the graces of declamation, and the natural gestures of the orator, the benefit will be enhanced. It is well known that Cicero, in early life, was predisposed to consumption; and Cuvier attributed his exemption from pulmonary disease, to which he was expected to fall a sacrifice, to the increased strength which his lungs acquired in the discharge of his duties as public lecturer.

Bathing, and "the art of swimming," should form a part of every boy's early education ; to the child predisposed to consumption, the daily ablution of the whole body is of the most essential service; it gives tone and vigor to the frame, frees the pores of the skin from those impurities which are constantly accumulating, and the muscular exertion which swimming demands is so universal, that not one part of the body is affected in a greater degree than another. A bath taken on rising, is most invigorating (see page 13); it preserves the body during the day at an equal temperature, and enables us to bear, with less risk of annoyance, any sudden change in the weather. Those who have never enjoyed this luxury, and have now the courage to commence, will not willingly lay it aside.

The period of life at which youth advances to adult age, termed puberty, extending in males from fifteen to eighteen, and in females, in our climate, from twelve, thirteen, or fourteen, to sixteen, is one of great importance to the future life of every individual; but important in an especial degree, to such as may be predisposed to consumption. Girls, especially, should not be over-crowded with studies, and should by no means be confined to schoolrooms, music lessons, evening studies, or any other arduous exertions of body or mind; but rather, have long vacations, with much exercise in the open air.

The selection of a proper occupation for a delicate or scrofulous youth, and the age at which he should commence the business of life, is an affair of no small importance. He should not be confined in crowded, heated, ill-ventilated factories; nor employed in any sedentary business, as that of a tailor, shoemaker, watchmaker, etc.; nor as a clerk at the desk, nor an engraver; he must not breathe an atmosphere loaded with irritating particles. At the same time, he should regard with care the rules we have already laid down under the *Laws of Health*, page 8; by which he may hope to earn for himself a new constitution.

We have, in practice, daily to combat the erroneous opinions of over-indulgent mothers, that a " delicate " girl is unable or unfit to walk, hop. or run, as her fancy may dictate ; and that she must be restrained in her movements, fettered in stays, and confined in a chamber warmed to fever heat. If it is wished that a delicate girl should become a sickly woman, such would be the plan to follow; but, if we desire to banish this delicacy and susceptibility, and give health and energy to the growing frame, we must allow Nature an opportunity of exerting her own powers.

In the early life of females, strict attention should be paid to the carriage, and the proper expansion of the chest; calisthenics are a useful auxiliary to health, insuring at the same time ease and grace of movement. In reference to this subject, Dr. Rush says, " Surely, it is not necessary, in order to acquire all the air and gracefulness of fashionable life, to banish from the hours of recreation the old rational amusements of battledore and shuttlecock, of tennis, trap-ball, or any other game that calls into action the bending as well as the extending muscles, gives firmness to every organ, and the glow of health to the entire surface." To prove the benefit of air and exercise, we have only to contrast the damp hair, the pallid features, and attenuated form of the young milliner, confined in a heated room for sixteen or eighteen hours, with the rosy tint and bloom of health of the more fortunate girl who is allowed to exercise in the open air.

Whilst guiding the physical education or " training" of a young person affected with a consumptive diathesis, we should not neglect the moral and intellectual culture. The passions now begin to exert a powerful influence on the health; it is now that the mind rushes into a new world, and is prone to receive lasting impressions, either of good or evil; new thoughts, new feelings, engage the attention; and the ideas and habits now acquired, whether amiable or vicious, frequently become a part of our future existence. Parents, especially, should not fail to caution children on the fearful effects of *solitary vice:* and bear in mind. that many a decline is so caused. See article on *Self-Abuse*. It is necessary that all gloomy and dispiriting ideas should be dispelled, and whatever tends to depress the mind or lower the animal spirits, should be avoided with the greatest circumspection.

The greatest discretion should be exerted in the selection of those who are to become the intimate companions of youth ; there are so many circumstances dependent on this choice, that materially affect the future health and well-being of the rising man, which every parent will readily comprehend, that they require only to be attended to, in order that their importance may be acknowledged.

Intemperance, excesses of all kinds, precocity, and all things that tend to induce nervous irritability and muscular debility, readily become the parent of consumption ; to those already predisposed to the disease, they frightfully hasten its development.

The climate most favorable to preventing or retarding the development of tuberculous consumption, is that which is of a mild, dry, and equable temperature ; hence a change of abode has been recommended in all ages to those whose native soil is subject to considerable and sudden variations.

Consumptives who go South in the autumn, usually go too late, and especially return too early in the spring. It is not so much the excessive cold of our climate which is to be avoided, as its great changeability. Iceland is remarkably free from consumption ; but then little or no pork is used there. There are no months in the year more trying than May and the beginning of June ; and yet invalids, unless strongly cautioned, make their arrangements to return in those months. Many a sick one whose disease has resisted the steady cold

of the winter months is cut off during the fresh
winds and cold storms, which alternate with the
hot days of spring and early summer.

In regard to a change of locality for consumptives, and the selection of a place for the few who
have the means and inclination to abandon the
comforts of home, California, with its warm and
equable climate, presents many attractions. The
sea-voyage, or overland journey thither, is no doubt
very beneficial in many cases. The very best locality is, probably, the Island of Santa Cruz. The
climate there is so warm and equable, that the
thermometer ranges but very little from eighty-two
degrees, the year round ; and the air is pure and
bracing. It is easily reached by steamer from
Havana. There is no denying that *very many*
cases of consumption are curable there, as late as
the beginning of the third stage, or even if large
cavities have formed ; together with the use of the
remedies already recommended.

Those who are able should return gradually from
the South homewards, stopping at St. Augustine, in
Florida, — which is *an excellent place* for consumptives to winter at, — and not reaching home before
July 1st.

A sea-voyage to any warmer climate is likely to
benefit consumptives. The exercise of sailing affords motion without exertion, or, at least, with no
more exertion than gives a pleasurable and tran-
quillizing feeling to the system ; it cheerfully engages
the mind, retards the pulse, calms the irregularities
of the heart, and produces sleep.

To prevent the ravages of consumption in one
already predisposed, especial attention must be paid
to bathing, nourishment, air, and exercise ; so that
he may be placed in circumstances the most favorable to acquire robust health. By removing func-

tional derangements as they occur, by maintaining a healthy condition of the digestive organs, and, above all, by obtaining prompt and efficient medical aid on the advent of the slightest pulmonary disturbance, we may confidently hope so to invigorate the constitution, as to turn aside and overcome the liability to tuberculous disease.

In regard to medical means, persons of consumptive family should be very watchful, and on the appearance of any symptoms or troubles for which Rush's Sarsaparilla and Iron (see page 47) is recommended, it should be promptly resorted to as the only sure medical preventive. It is peculiarly adapted to the youthful organism, and should be made use of by all persons of consumptive family during the period of approaching womanhood and manhood.

MEDICAL HUMBUGS.

The different modes in which consumptives are defrauded, and the public deceived in medica' matters, are here exposed.

Consumptives are more exposed to different swindles and impositions than almost any other class of persons, for the reason that their supposed hopeless condition leads them to grasp at any means of relief offered, which seems, to them, to promise any plausible chance of success. Various are the modes in which ingenious but dishonest men have enriched themselves, and plundered their victims, not only of their money, but of precious time, during which their disease has made such progress as to become quite incurable. The first of these, in point of frequency, consists in a person who is not a medical man announcing himself as such; or falsely pretending to be a clergyman, in a short advertisement, and offering to send, free of charge, a wonderful prescription, by which *himself*, or *a favorite daughter*, or some one else, who was given up to die of consumption, realized a most wonderful cure. Such was the celebrated " Retired Physician," whose " sands of life had nearly run out ; " and such, we are informed and believe, is now a person who claims to be a clergyman, and advertises from Williamsburgh, N. Y. The prescriptions sent by these impostors contain ingredients wholly unknown to any medical man, or at any drug-store ; and which, of course, cannot be procured without writing to *them*. Some three to five dollars is then demanded for an utterly worthless, and often injurious combination of drugs, which would not be taken by the sufferer if he knew their *true* names ; and which would not cost ten cents at any respectable druggist's. The principal ingredient in the " Retired Physician's " receipt was the extract or tincture of Indian hemp, which is also known under the name of *cannabis, hashisch,* or *hasheesh ;* it affords some relief at first, but like opium and morphine, proves, ultimately, pernicious to the lungs. But this prescription put many thousand dollars into a notorious swindler's pocket, because the drug could not be found at most drug stores ; and the poor consumptives sent to the " Retired Physician " for it. The other person alluded to (the Rev. impostor, at Williamsburgh, N. Y.), on being written to, announces in a pamphlet that *he himself* has been cured of consumption by a receipt, which he also sends, and which reads as follows : —

" Extract of Blodgett, three ounces ; hypophosphites of lime and soda, one half ounce ; alantin, one drachm ; meconin, one half scruple ; extract of cinchona, two drachms ; loaf sugar, one pound ; port wine, one half pint ; warm water, one quart."

It will be seen that the *Extract* of Blodgett is one of the principal ingredients ; and *this* can be procured *only* of him. There is no medicine by that name ; and the entire prescription is a shameless imposition.

The *name* (extract of Blodgett) is so laughable, that he goes on to say, in his pamphlet, by way of explanation, that it is shortened from the Italian Blodgetti, and known in the works of Italian authors. The editor of this book, when in Rome, Naples, Milan, and Venice, incidentally made inquiry, and satisfied himself that it is also totally unknown there. *Port wine* is *always* more or less adulterated, and is an injurious article for any sick person.

The Respirometer Humbug.

A respirometer is an instrument to measure the amount of air breathed into the lungs, or out of them, at one time. It is *practically valueless in lung complaints ;* for the reason that this amount of air varies so much in healthy persons, that it is quite double in some persons to what it is in others. We consider the use of the Respirometer, therefore, *as entirely and solely a device, or manœuvre, to extract money from the pockets of the unfortunate.*

The Oxygenated Inhalation Humbug.

This is one of the very latest. The gas used, however, is not oxygen, but merely nitrous oxide, or laughing gas. It is merely an unnatural stimulant to the lungs, somewhat like chloroform, and if enough is taken would produce intoxication. It never does any more than temporary good, and may do permanent injury. It is only another device to fleece the unfortunate.

The Speciality Humbug.

It is usually the perfection of mercenary imposture for any man to claim to know more of ONE disease, or the diseased action of any ONE organ, or any one class of organs, because he CONFINES his treatment to one only! Such a man is sure to make every case belong to that class to which he confines his treatment, if it bears the least resemblance to it, that thus he may secure the patient; and, what is of more consequence in his eye, viz., his fee. Moreover, having to do with one class of maladies, or the symptoms of a single class, he loses sight wholly of all the other maladies which have preceded the kind he treats, and which are the hidden causes of all the mischief; hence he fails to detect the case as it really is, treats symptoms merely, and must end in disappointing, if not ruining his patient We have no patience with such wretched charlatans who torture the feeble and the sick with their false and futile treatment, doing infinitely more harm than it is in the power of the best physicians to remove. To be FAMILIAR WITH ALL DISEASES, and the numerous and varied causes of every form of sickness, we must be DAILY IN THE HABIT OF EXAMINING ALL SORTS OF CASES, and TREATING ALL, thus we keep every symptom, every source, every appearance, every change of disease, of all the various organs, FRESH IN OUR MINDS; hence are ever prepared to meet any case, and to understand it fully; and, be it what it may, to represent to the patient what it really is, and to treat it according to its real character. As universal practitioners, we are never tempted to pronounce every affection of the chest *Consumption;* to make every derangement of the uterine functions the *primary* or the *principal* ailment, when such derangements are nothing more than symptoms, or merely parts of a complicated disease ; to make diseases of the *eyes*, or *ears*, or *limbs*, or *glands*, local, so as to justify a barbarous surgical operation ; nor to make all affections of the genital organs gonorrhœal, or syphilitic, as those smutty venereal quacks do who confine all their quacking, cobbling, and poisoning to this class of diseases alone.

The mercenary character of such men is also one of their chief characteristics. There is at the present time such a person who comes from a distant city, and whose pretentious advertisements fill whole columns of the newspapers. The editor is informed that his charges in his Speciality of Consumption are enormous. It is quite evident that in cases of such very large advertising, *somebody* must make the money to pay for it.

The Old School, or Calomel Doctor Humbug.

It is rather an ungracious duty to speak thus of a large class of educated practitioners. But in what other terms can truth be told in regard to their treatment of such a disease as consumption ? Do they not, and country physicians especially, pronounce the disease incurable, and does not the result of their practice prove their assertion true ?

It is next to *sure death* for a consumptive to fall into their hands ; and his executors will not fail to find a remunerative claim against his or her estate, or friends, in the shape of the *doctor's bill*, for practice according to Gunter. The patient, however, has the advantage of dying according to law and custom. We will, moreover, add that the Thompsonian, or Botanic doctor, is not a whit more successful, on the average, than the so-called regulars. Drenching the system with quarts and gallons of drugs is no way to cure consumption, or any other disease

The Indian Doctor Humbug.

Of all the follies in medicine, none exceeds that most absurd one which places reliance on what are called INDIAN DOCTORS — most of whom are impostors, who know no more even of the Indians than they do of medical science or the human constitution. But what do INDIANS know of medicine? Comparatively, nothing. What do they know of the structure, and organs of the human economy; the lesions to which they are liable; or any other branch of true medical philosophy? Nothing. They may cure a cold, a slight fever, a simple sore or ulcer; and so can many old and observing nursing women do the same. They know the medical virtues of but few roots and herbs, and they make no progress in knowledge, because they have no facilities or aids for doing this. When any violent disease breaks out among them, every soul perishes; or they flee in all directions, usually to white settlements, for relief or safety.

Accidents. Concussions. Falls.

A person injured by a fall, by the shock, or concussion, and taken up insensible, should *not* be bled. Loosen his collar, and waistband, and carry him carefully home or to the nearest house, put him in bed, and make frictions with warm flannels, and apply mustard plasters to the feet and calves of the legs ; these should not be left on to blister, but only to produce redness. As soon as he is able to swallow give hot rum or whiskey toddy by the tablespoonful. A bottle of tincture of Arnica should be kept by every family, a receipt for making which may be found in this book on page 151. It is very valuable for all injuries, bruises, wounds and accidents; and may be used as a liniment, and for internal administration. In such a case as the above, put three drops in a tumblerful of water, and give a tablespoonful every hour. Also rub any bruised place with a little of the tincture.

Drowning.

When a body has been removed from the water, it should be removed to the nearest house and placed in a warm bath, if it can be had, of the temperature of 100 degrees ; if not, place it in a warm bed, and keep up the friction by means of hot dry flannels. Bottles of hot water, or hot bricks wrapped up in moist cloths should be kept at the feet and sides, taking care that the bricks are not too hot *But the chief remedy is artificial respiration.* This may be done by placing the patient on his back, and pressing out the air from the lungs *very strongly*, by pressing on the chest; then turn him on to his side which will cause the lungs to fill again. If that does not seem to succeed, the lungs may be inflated with common bellows. About 12 inflations a minute are sufficient. Artificial respiration should be continued *several hours*; especially if the person has not been long in the water. At the same time continue the warm applications.

Hanging.

Persons found hanging should be immediately cut down with care : and treated in much the same way.

Wounds.

If a wound is bleeding in a steady flow, and dark colored blood, the edges should be carefully brought together with the hand, and a large piece of cotton, or lint, firmly bound on with a bandage; or what would be better, use long strips of sticking plaster cut about half an inch wide, and draw the edges of the wound close together in applying it. If the flow of blood is in jets, and of a bright red, it is an artery that bleeds, and requires tying. Until a surgeon arrives, tie a strong handkerchief firmly round the limb, between the wound, and the body, and twist the handkerchief with a stick, until the bleeding stops.

The course of the main artery is at about the middle of the inside of the limb; if a knot is tied in the handkerchief, and applied over the artery before the handkerchief is twisted, that will help stop the flow of blood. This ligature may be fastened in place until the artery can be tied.

Burns and Scalds.

These are best dressed by sifting on flour, very abundantly, from a common flour box; the benefit of which is to exclude the air.

Where burns are extensive, and there is much prostration from the pain, give a little hot currant wine, or rum toddy to bring on reaction. If the pain is very severe, cloths wet with cold water may be applied over the thick coating of flour. These will give relief if kept wet and cold. A burnt finger is at once relieved of the pain by putting it in cold water.

Frost Bites.

A frozen part becomes white, and may mortify if thawed suddenly. Snow should generally be applied, while the circulation is being restored and no rubbing should be used, lest the frozen cells be broken.

A person benumbed with cold, and nearly, or quite insensible, should be taken into a cold room, and be rubbed with snow first, and flannels afterwards, until improvement takes place. Warm wine and water may be given as soon as it can be swallowed.

Useful Medicines.

Rush's Sarsaparilla and Iron, page 47.
Rush's Monthly Remedy, page 231.
Rush's Pills, page 63.
Rush's Lung Balm, page 54,
Rush's Restorer and Preventive, page 231.

The proof spirit mentioned in the following medicines, should be half-strong alcohol, and half water.

Aconite. Mix one ounce proof spirit with ⅛ ounce tincture of Aconite, (root or plant).

Arnica tincture. Put one ounce of Arnica flowers in eight ounces of proof spirit. It will be fit to use in a week.

Arnica Plaster. Take flesh colored court plaster and brush it over with tincture of Arnica, and then let it dry. Or take the new kind of sticking plaster, which sticks by wetting, and brush it over in the same way. This is very good for corns.

Belladonna. Mix ⅛ ounce of tincture of Belladonna with one ounce of proof spirit.

Chamomile or Chamomilla. Mix ⅛ ounce of tincture of Chamomilla, with one ounce of proof spirit.

Coffea. Put three kernels of green coffee, cut fine, into an ounce of proof spirit. It will be ready to use in four days. It is very good for restlessness.

Hamamelis. Mix ⅛ ounce of tincture of Hamamelis, with one ounce of proof spirit

Pulsatilla. Mix ⅛ ounce of tincture of Pulsatilla, with one ounce of proof spirit.

Rhus tox. Mix ⅛ ounce of tincture of Rush tox, with one ounce of proof spirit.

Ipecac. Mix ⅛ ounce of tincture or wine of Ipecac, with one ounce of proof spirit.

When the above medicines cannot be obtained at common drug stores, they can be had at the Homœopathic, or Botanic. The directions for using them are given under the appropriate diseases. The following are much used.

Wine of Ipecac. Dose as an emetic for an adult, two tablespoonfuls; as an emetic for a child one or two years old, one teaspoonful, repeated every fifteen minutes till it operates.

Hive Syrup. Dose for a child of one year 15 drops; from six to ten years, 20 to 40 drops; it is mostly used for croup.

Paregoric. Dose for an adult, one to two teaspoonfuls; for a child one year old, the dose is from three to ten drops.

Laudanum. Dose for an adult, fifteen to thirty drops. It should not be given to children.

Poisons.

The first thing to be done, in case of poisoning, is to give an emetic, and this may always be effected by giving a tablespoonful of mustard, mixed in a half pint of warm water. After vomiting begins, give more warm water to promote it; copious vomiting will dislodge most vegetable and mineral poisons.

The acids generaly are antidoted by soda, saleratus, or soap suds; *Sugar of Lead*, by Epsom Salts; *Corrosive Sublimate*, or bed bug poison by white of egg, and milk; *Verdigris and Copper*, by white of egg, and brown sugar.

Opium, Morphine, Laudanum, and Paregoric; as well as *Alcohol* and whiskey, which may poison children; require abundant cold bathing of the head, or whole body. Pour the water on the head by pitcherfuls, and in bad cases, keep the patient in motion, by walking him round. Strong Coffee, may also be given, with advantage; give two tablespoonfuls very strong every fifteen minutes, without sugar, or milk.

Arsenic requires a prompt emetic of mustard; and when it can be had, the *hydrated peroxid of Iron.* Give also the white of 10 or 12 eggs beat up in a quart of warm water, with the mustard. When the *oxid of Iron* cannot be had, if you have *copperas*, and soda for bread, dissolve a heaping tablespoonful of each in a pint of warm water, and pour them together in a large dish; they will foam; let this subside, and then give a teacupful of this at a time stirred up with a heaping teaspoonful of mustard, to ensure vomiting.

FEVERS AND INFLAMMATIONS.

In most acute diseases fever is present. The symptoms common to most fevers are ;—at first, a chill, or chilliness, and shivering ; and then heat, thirst, restlessness, prostration of strength, and usually headache. When there is an inflammation of a part, there is usually pain in that part. The symptoms, or signs of inflammation, besides *heat* and *pain*, are *redness and swelling*. When an inflammation is on the surface of the body, all these signs may be noticed ; take for instance, *a boil ;* or *erysipelas*. But in case of an internal inflammation, the *redness*, and *swelling*, can not usually be seen. Take for instance : *Inflammation of the lungs*, commonly called *lung fever*. Here the *heat*, and *pain*, are both very sensible ; but the *redness* and *swelling*, being internal, are not visible.

Physicians do not mix up *fevers* with *inflammations ;* but in common language most inflammations are called fevers . for instance, *inflammation of the pleura*, or membrane which covers the lungs, and inside of the chest, is called *pleurisy fever*. It will be well for the reader to bear this in mind, and also the fact that *fevers possess the property of passing from one kind into another. Remedies and treatment, therefore, should always be adapted to the symptoms of the case,* and given on general principles; not for the *name* of a disease. Many fevers have what is called *a turn* ; that is, they suddenly give way, and the patient improves from that time ; but this is not always the case. Many such attacks decline gradually, and have no *turn* or *crisis*.

The causes of fevers are numerous. Miasms, epidemic influences, contagion, powerful mental emotions, derangement of some important organ, external injuries, excess, or errors in diet, heat or cold, or alterations of temperature, exposure to cold or damp, in fact, anything that causes derangement of the equilibrium of the system, may produce fever.

GENERAL TREATMENT IN FEVER, AND DIET.

The great essentials in the treatment of fever are :

Perfect rest, mental and bodily.

Pure air, thorough ventilation, and a cool apartment; the temperature of the patient's room should never exceed 65 degrees.

Feather-beds should be discarded, and mattresses substituted, when practicable.

Nature herself generally prescribes the regimen to be observed by taking away appetite ; while the thirst present, as Dr. Rush has well observed, may be considered as her voice calling for fluid. Water is the best diluent ; drink a plenty of it ; no solid food, broth, or even gruel and the like, should be permitted in cases where the inflammation runs excessively-high : and the utmost caution is to be observed, in allowing gruel or weak broths during the decrease ; an error in this respect often causes irreparable mischief, and it is always safer to err a little on the side of abstinence than on that of indulgence.

A little toast-water, or weak barley or rice-water, sweetened with a little sugar, raspberry or strawberry syrup, may be allowed when the fever is somewhat abated, though then we must still carefully avoid incurring the risk of a relapse, by giving any aliment likely to tax, in however slight a degree, the digestive powers.

Simple Fever from taking cold usually lasts but about 24 hours ; it may be known as we have stated above by *a chill,* and *shivering, heat of skin, restlessness, rapid pulse, usually headache, and slight loss of strength.* This may occur in children as well as adults. In such a case it would be well to put two teaspoonfuls of *Rush's Lung Balm* (see page 54) into a tumblerful of water, and give an adult a tablespoonful once in an hour and a half. A child may take a teaspoonful. This will usually bring on a perspiration, which will terminate the attack Every attack of simple fever may be usually thrown off by the use of this remedy in about 12 hours. It will never do any harm to use it, in case it proves a severe fever; and will certainly prevent it from settling on the lungs. *In all fevers the feet should be kept thoroughly warm with bottles of hot water, or warm bricks.* The best way perhaps is to use bricks. Put two or three at the fire, and on trying them you will usually find one side too hot: cool it by pouring on water; then wrap it up in a cloth, not paper, and a *moist* heat will be given out. Keeping the feet warm, draws the blood away from vital organs, and equalizes the circulation. It is of most importance, between midnight and morning; and patients with fevers frequently take cold then.

Inflammatory Fever.

This name is given to an attack more violent than the above, which may last, under common treatment a week or two. Children are not subject to it. The symptoms are more violent than in *simple fever:* the pulse is strong, and hard; the tongue is coated white, or else of a brighter red than natural; the urine is red, and scanty; there is constipation; the breathing is hurried; there is usu-

ally severe headache ; the chill, heat, and dryness
of skin, and prostration of strength, are more severe than in *simple fever*.   A person taken sick in
this way should take Rush's Lung Balm without
fail.   Put two teaspoonfuls in a tumbler of water,
and give a tablespoonful.   Put five drops of Aconite (see page 151,) in a tumbler of water, and
give a tablespoonful in one hour after the first
dose.   Continue these in that way alternately,
every hour, until the patient perspires freely ;
then leave off the Aconite, and give only the
Lung Balm once in two or three hours, until the
patient is entirely relieved.   But if the fever rises
again, resume the Aconite, and Balm, both.   If
the bowels are constipated give three of Rush's
Pills, and repeat the dose every six hours, until
the bowels are moved.   *Be sure and keep the feet
warm* as directed under simple fever.   If the patient is used to daily bathing he should not fail to
go into a *wet sheet half-pack*, or full-pack, as directed under water treatment, as soon as the attack comes on.   Let the wrapping outside be light ;
and let the pack be opened, and the sheet be wet
again, and again, as fast as it gets warm, until the
fever is subdued.   This may be known by the feeling of coolness, and comfort ; and perhaps the patient falls into a sweet sleep.   It usually takes from
three to twelve hours to effect this, and in severe
attacks may be repeated for three or four days in
succession ; if the fever rises again and again, as
it may do.   The wet sheet pack will usually get
heated up in about half an hour ; it should not be
allowed to get hot and dry.   Keep it cool by
opening it, and wetting the sheet, and give the
medicines as directed, and you will surely kill the
fever.   Put a wet towel on the forehead, and keep

it well wet and cool, for the headache. The Aconite should be given while the patient is in the pack, as well as the Lung Balm; as before directed; the Balm *only*, after the pack is over. If the patient is a delicate person; is not used to water; or afraid of the pack; give the medicine only; and sponge the chest, back, and body all over, every two hours or oftener, with a basin of cold water. Cases of most violent fever, which sometimes prove fatal, or last two, three, or four, or more weeks, rarely last more than three or four days when treated with these medicines, and water both. After taking a pack as above directed, and the heat of the skin, and the feverishness, being for the time subdued; let the patient be well sponged over with cold water, or stand in a tub and be washed down with a basin of water, and towel, or sponge, and then be put into a clean bed. No one can know the comfort, nay, luxury of such use of water until experience of it is had. The writer of this knows of what he speaks. having recovered from a most violent attack of fever in three or four days under just such treatment as this. Yet those who are afraid of water, may rely upon RUSH'S LUNG BALM, and *aconite* alone, as superior to any other remedies, for attacks of inflammatory fever. It is the nature of this kind of fever to be worse at night, or from noon till midnight; therefore after the feverish action has been once subdued, we should not be disappointed if we are obliged to renew the treatment as at first, in the afternoon, or toward night. As soon as the patient is convalescent, and is able to bear considerable nourishment, some weakness and debility may, and probably will remain. In that case Rush's Sarsaparilla and Iron should be taken, and will completely renovate the system. One bottle will usually be sufficient. Many a decline which has come on, after violent fevers, might have been prevented by taking a bottle of this medicine.

Typhus and Typhoid Fevers,

Usually begin very much as already described *as simple, and inflammatory.* They may always be safely treated by RUSH'S LUNG BALM, *Aconite,* and sponging with water, for three or four days; when, if the attack does not seem to give way, a physician should be sent for. After a long and debilitating typhoid fever, Rush's Sarsaparilla and Iron should invariably be used, to invigorate, and restore the system. It should be continued until it ceases to benefit; or the usual health, and strength is restored. In some cases, one bottle will suffice. In other cases, two, four, or six, will not be too much.

Fever and Ague, or Intermittent Fever.

A severe chill, followed by burning fever, and this, by profuse perspiration, denotes the presence of this disease. The treatment is as follows. As soon as the chill comes on, put ten drops of Aconite (See page 151,) in a tumbler of water, and give a tablespoonful every ten minutes while the chill lasts. Continue it every half hour during the hot stage, or while the fever lasts. As soon as the sweating begins, leave off the Aconite, and give the following preparation. Put forty-five grains of quinine into one half bottle of Rush's Sarsaparilla and Iron; mix it, and shake it well. Take a tea-

spoonful every three hours, until the chill returns.
Then give the Aconite as before, and so on.
There will generally be but one chill after you
commence; and very rarely more than two. After the chill is broken, continue the Sarsaparilla
and Iron and Quinine, by giving thirty drops every six hours until the mixture is finished. Then
go on with the remaining half bottle of Sarsaparilla and Iron, and take it according to the directions on the bottle. This mixture of Sarsaparilla,
Iron, and Quinine, never injures the system; but
will restore even broken down constitutions. After being cured of the Ague, if debility remains,
continue the Sarsaparilla and Iron until the health
is quite restored. If the bowels are at any time
constipated, they should be moved by Rush's Pills.
These may be taken at bed time.

Inflammation of the Brain. Brain Fever.

This is fortunately a somewhat rare disease.
It usually results from irritating causes; such as
mental emotion, or excessive mental labor; ex-
tremes of heat, and cold; injuries; concussions,
and excesses. It usually begins with the symptoms
of *Inflammatory Fever* (which see) but the head
becomes affected very soon in the disease. We
have great heat of the head; boring of the head
in the pillow, with violent pulsations of its blood
vessels, and of those of the neck: there is extreme
sensibility to noise, and light; the face and eyes
are red and swollen; the pain in the head is violent, shooting, and burning; there is furious delirium. An adult person seized with these symptoms should be treated promptly as follows. Put

five drops of Aconite in a tumbler of water and give a tablespoonful:—in one hour after put five drops of Belladonna in a tumbler of water, and give a tablespoonful, and so go on with them every hour alternately. Keep the feet well warm, as directed under simple fever. The head should be constantly sponged, and sopped with water from a large basin with ice in it; keep the hair sopped full of the ice cold water, and let the water soak into a sheet, or a straw pillow, which can be changed often enough to keep from wetting the bed. Wet the back part of the head as well as the front. If the bed gets wet it will do no harm, *if the feet are kept warm*, and the patient is put into a dry bed, as soon as the fever gives way; when that takes place give only the Belladonna every two hours. If the fever, when once subdued, returns; go back to the same treatment, as at first. If the bowels are constipated give four of Rush's Pills, and repeat the dose if there is no motion in four hours. In children, this fever may be treated in the same way, giving a *teaspoonful* of the medicine for a dose, instead of a tablespoonful. Attacks of Brain Fever may usually be thrown off in this way, in a day or two if treated promptly. If not no harm will have been done, if a physician has to be sent for.

Inflammation of the Lungs. Lung Fever. Pneumonia.

This is one of the most common diseases, both for adults and children. It commences with fever of the inflammatory type, (see inflammatory fever,) and, in addition to the symptoms of that disease, we shall find a short, dry, and continual cough, and difficulty of breathing; with expectoration of tough mucus, or phlegm; at first whitish, but after a day, or two, of a *rusty color*, which is considered, when present, a sure sign of lung fever; but it is not always seen. There is generally some pain in the chest, which may serve to point out the seat of the disease. The treatment of the Lung fever is to be, precisely the same as for Inflammatory Fever, (see page 155) except that we should be sure to get the feet well warm before applying the half-pack, or sponging. Put the feet in hot water at once, if cold, and commence with the medicines; (Balm and Aconite.) After the feet are once warm, keep them so with hot bricks. The half-pack may then be applied, under the same conditions as in Inflammatory Fever, and the medicine given the same. A child should take a teaspoonful; and a very young child, half a teaspoonful of the mixture, for a dose. It is better to use the towel-pack round the chest of a child, by wetting a towel in cool water, squeezing it out so as not to drip, pinning it round the child's body, up to the arm pits, and then putting a dry towel round it all. This should be renewed, by wetting it again, as soon as it begins to get hot. Do not let it get hot and dry; as it would then do harm instead of good. A pack in this way will never harm even a child, *if the feet are kept warm.* A *Lung Fever* rarely runs more than five days, and is frequently broken up in a day or two, when treated in this way.

11

Inflammation of the Pleura, Pleurisy Fever.

This disease may be known by a sharp, sticking pain in the side; burning fever, like lung fever; not much cough; difficult breathing, but less oppressed than in lung fever. The patient lies generally on his back, or the painful side. The nature of the disease consists in an inflammation of the membrane which covers the outside of the lungs, and the inside of the chest. Hence the sharp pain when they rub together; the two surfaces being naturally in contact. The treatment is to be precisely the same with *Inflammatory Fever and Lung Fever*; which see.

Inflammation of the air passages.
Bronchitis,—Ministers' Sore Throat.

The symptoms very much resemble Lung Fever, but are more mild, and like a common cold. There is *hoarseness*, and much wheezing, and cough, with expectoration of phlegm, and the breathing is quick, and oppressed. In children, phlegm frequently impedes the breathing, and vomiting ensues. Fever is sometimes mostly absent; and those are apt to be dangerous cases; because not attended to. *In any case like the above*, give small doses of Rush's Lung Balm every three hours until relief takes place. An adult may take 20 drops; a child less, according to directions on bottle. If the throat is hot, and feverish, you may put a wet towel round it, with a dry one outside, and keep it cool, by wetting it again as soon as it begins to get hot; the same as for *croup*. If there is heat of skin, and high fever, give aconite, one dose, and

Lung Balm an hour after; the same as directed for *Inflammatory Fever*, (which see) and keep on alternately in that way, until the fever abates. This disease is frequently chronic, and sometimes called *Minister's Sore Throat*; it may be invariably cured by the Lung Balm, taken according to directions.

Inflammation of the Liver. Liver Complaint. Jaundice.

This is more common in hot climates, than here; and very many cases of disease, where the stomach and bowels are at fault are saddled upon the liver; and mercury in some form given; which, though it affords temporary relief, ruins the system; and has to be resorted to, more and more frequently.

An attack of Liver Complaint may be known by pain in the right side, and usually some swelling over the liver; tenderness on pressure there; high-colored urine; yellow tinge of skin and eyes; moist cough; and symptoms of indigestion. In Jaundice the yellow color is more decided all over the body; the stools hard, and whitish. Never take calomel, or blue pills for this disease. Keep the bowels well open with Rush's Pills; of which a good dose, to open the bowels, should be taken at first; and then one every other night, until improvement is decided. At the same time take Rush's Sarsaparilla and Iron, a half a teaspoonful every four hours, commencing on rising. When the bottle is half finished a grain of Quinine may be added to the bottle, with good effect · to be taken the same as before. A bad case may require more than one bottle. Persevere until you are cured. The cure will usually be hastened by *a*

wet sheet pack, taken about the middle of the fore-
noon, or afternoon, one hour ; and a *Sitting-Bath*,
for a half an hour, before retiring.   See Water
Treatment, page 27.

Bilious Complaints

Of whatever nature, may be treated exactly as
above.

Dyspepsia. Indigestion.

This is too common to require much description.
The causes are very numerous ;—the most common
are ; intemperance in eating, or drinking ; eating
too much : use of coffee, tobacco, and spirits ; imperfect chewing of the food ; impure or hard water ;
sedentary and vicious habits ; and exhaustion of
body, or mind.   The signs of dyspepsia are briefly : —sick-headache, dizziness, and confusion of
head ; face pale, or yellowish ; tongue foul and
coated, dry, white, or yellowish ; bad taste in the
mouth ; *want of appetite* ; sickness at the stomach ;
bitter belchings, and vomitings ; *heartburn* ; *sour
stomach* ; *flatulence* ; fulness at the pit of the stomach ; clothes at the waist feel tight ; cramps at or
in the stomach ; red urine, with brick dust sediment ; sleep unrefreshing, and restless ; *constipation.*

Persons suffering with any of these symptoms
should have the bowels moved by Rush's Pills ;
and then take Rush's Sarsaparilla and Iron, after
each meal, and on going to bed.   At the same
time remove the causes, by living according to the
laws of health.   Take a sponge-bath every morning.   If there is *constipation*, or the above symptoms

well-marked; eat coarse, or Graham bread, or made of *wheat meal*; and *nothing* made of fine flour; only a little meat once a day, and *no swine's flesh*; a plenty of vegetables and fruit, such as agree with the stomach. Read the articles on *diet*, and *means of preserving health.* Coffee, green tea, tobacco, and spirits should be entirely abandoned. The blood is generally poor, and deficient in quantity, Rush's Sarsaparilla and Iron in all such cases will *do good*; and a complete restoration to health may be expected if bad habits are abandoned, and the laws of health lived up to. Persevere with the medicine, and do the best you can about diet, habits, &c.

Hæmorrhoids. Piles. Constipaton. Costiveness.

These troubles are usually brought on by the same causes with Dyspepsia, (see the preceding article,) and require about the same diet, and treatment. For constipation do exactly as there directed. *Coffee must be given up.* A tumbler of pure soft water taken at bed time is no contemptible remedy. A sitting bath for a half an hour before going to bed will well repay the time and trouble. See *Water Treatment* page 27. For Piles it is a valuable remedy. Coffee in many cases *causes* the piles; attend strictly to the rules under diet. For this complaint take a ripe horsechestnut, and grate half the meat, with a fine grater, and put it into a bottle of Rush's Sarsaparilla and Iron; let it settle, and take a half a teaspoonful after each meal, and at bed time. If you cannot get the horse-chestnut, take the medicine without. For Constipation, use the same remedy with-

out the horse-chestnut. Rush's Pills may be used
for cases not of long standing; and are always safe
and efficient.

Colic.

This may be *flatulent*, or *bilious*. The pain is
generally about the navel ; and is relieved by pressure ; by which we distinguish it from dysentery, or
inflammation ; which is made *worse* by pressure.
The common remedy of strong, hot peppirment
may be tried. If that fails, two or three teaspoonfuls of *paregoric* may be given once an hour until relief is obtained. A child must take less. Persons who are subject to attacks of colic may be
permanently cured by Rush's Sarsaparilla and
Iron.

Diarrhœa, Summer Complaint, Dysentery.

The causes, and signs of diarrhœa, summer complaint, and looseness of the bowels, are too well
known to require enumeration. Most cases result
from irritating or undigested food. It is therefore
a good plan to clear the bowels at first by a dose
of Rush's Pills; and then a few doses of Rush's
Sarsaparilla and Iron, a teaspoonful every three
hours, will complete the cure. If you have Blackberry Cordial (see Receipts) give each dose of
the Sarsaparilla in two tablespoonfuls of Cordial.
If it is the season of blackberries, get some blackberries, and make the syrup. It is a good remedy ;—to children give less in proportion to age.
you may give *paregoric* instead of Blackberry cordial ;—dose, two teaspoonfuls ; for a child less.

Dysentery generally begins with looseness of the bowels, but soon grows worse, there being inflammation of the lower bowels. This is attended with some fever and heat, great pain and straining at stool, with green, slimy, and often bloody motions. It may be treated for the first day or two the same as *diarrhœa*; but if it has reached the stage above described, which will rarely be the case if treated as above, domestic treatment should be no longer attempted. You may, however, apply a towel wet with cold water over the bowels, and keep it cold by changing it. Also throw up an injection of cold water, as large as can be retained five or ten minutes, say a pint or more, *after each motion of the bowels.* Keep the feet warm. All this will be very grateful to the patient, and will often succeed in a few hours in arresting the disease. It cannot do any harm, and will not interfere with any other medicine. After a debilitating sickness with dysentery, it is highly important to take a bottle or more of Rush's Sarsaparilla and Iron, to restore the blood, and system generally. The diarrhœas and dysenteries of children may be cut short by Chamomilla; (see page 151) put five drops of it in a half tumbler of water, and give a teaspoonful after every loose motion. Give first, to a young child, a little tincture of Rhubarb, to clear out the bowels; then the cordial, paregoric, or Chamomilla, which may be tried one after the other.

Cholera Morbus. Cholera.

The first named is liable to occur in hot weather, and sometimes at other times of the year. It commences with nausea, and griping, followed by vom-

iting and purging.   In severe cases there will be
coldness of the body, and hands and feet; great
thirst; cramps in the belly and legs; shrunk features; pulse very weak; discharges thin, watery
and fetid, with strainings and bilious vomiting.   Attacks of this kind are sometimes, but not always,
preceded by diarrhœa.   This should be promptly
checked as directed under *diarrhœa*; *especially in
a cholera season*.   If the skin, and hands, and feet
are cold; give tincture of Camphor, of which put
two teaspoonfuls in a cup of warm water, and give
a teaspoonful every five minutes.   At the same
time apply hot bricks, wrapped up in damp cloths,
to the feet, calves of the legs, sides, and arms;
and continue the treatment till reaction takes place,
and warmth is restored.   When this has taken
place, if vomiting continues, put a tablespoonful
of Rush's Lung Balm in a teacupful of water, and
give a teaspoonful after each attempt to vomit.
After the vomiting has stopped, the remaining diarrhœa may be treated as directed under *diarrhœa*.

In real Cholera cases the treatment with Camphor should be exactly as above, until a physician
arrives.   Experience has shown the Homœopathic
treatment to be the most successful for Asiatic
Cholera.   A person recovering from Cholera Mor-
bus, or Cholera, should take Rush's Sarsaparilla
and Iron; to effect a complete restoration of the
system.

Rheumatism.

This disease is most generally *chronic*; and affects the joints, and back; and muscular, and
membranous parts.   The signs are stiffness, and

tenderness, with some swelling and numbness. It is caused by exposure to cold, and dampness, and other depressing causes. It may be very successfully treated as follows. First clear the bowels if constipated, by Rush's Pills; then get a bottle of Rush's Sarsaparilla and Iron, to which add one ounce of Wine of Colchicum, (root or seed); take a teaspoonful after each meal, and at bed time. At the same time take a vapor bath, or some kind of a sweat, (see page 31,) every two or three days, so as to perspire well. This will greatly assist the action of the medicine. Very few cases of Rheumatism will resist this treatment. In cases where it chiefly affects the joints, or small of the back, constituting *Lumbago*; entire cure may be experienced from Rush's Rheumatic Plaster, which may be easily procured by writing directly to the editor; address at page 44. This plaster will at once relieve the most troublesome pains, and aches; and for those who are much exposed, and cannot afford to lose time, it is the best remedy.

Acute Rheumatism. Rheumatic Fever.

'This is attended with fever of the inflammatory type; (see Inflammatory Fever,) with the usual signs of rheumatism, in a more severe form than first spoken of. The pains in the large joints often shift about, and leave redness, swelling, and tenderness; and much weakness. If there is feverishness, put five drops of Aconite (see page 151,) in a tumblerful of water; put thirty drops of Wine of Colchicum in another tumblerful; put a tablespoonful of Rush's Lung Balm in another tumblerful; give a tablespoonful of the Aconite first; in one hour give a tablespoonful of the Colchicum; and after

another hour give the same of the Balm ; and so
go round with the three. As soon as the fever
abates give only the Colchicum, and the Balm ;
with an hour and a half between the doses. The
fever having subsided, give a good sweat with
blankets, and hot bricks, which will finish breaking up the disease. To complete the cure, take
the medicine as directed for Rheumatism. Joints
remaining painful should be packed at night, and
during the day time, if confined to the house, with
a wet towel, with a dry towel, or flannel round it.
This may be applied to a swollen, and painful
knee, at any time during an attack, with advantage.

Tic Doloureux. Face Ache. Sciatica.
Neuralgia.

These are all nerve-pains. In the first named,
or *face ache*, we find shooting, and darting, cutting,
and tensive, and convulsive tearing and dragging
pains, about the eye, face below the eye, and tem-
ple. Great relief will frequently be found from
Belladonna ; of which put five drops in a glass of
water, and take a teaspoonful every hour. Permanent cure will generally be found for this complaint,
for Sciatica, and for all other kinds of Neuralgia,
by taking Rush's Sarsaparilla and Iron ; which
strengthens the whole system, and enables it to
throw off the disease. We recognize *Sciatica*, by
pain in the region of the hip joint, extending to the
knee, and foot ; and following the course of the
Sciatic nerve. It often interferes with the motion
of the foot, causing stiffness. The remedy, without doubt, is Rush's Sarsaparilla and Iron, (see
page 47). Persevere with it.

Headache.

This is a species of neuralgia, and there are many kinds of it. One medicine will not cure all kinds of headache. *Belladonna*, (see page 151) will cure more headaches, than any one kind of medicine. The symptoms which indicate it are:— feeling of fulness, and pressing and expansive pain as if the head would burst open; pain in the forehead; about the eyes; or extending over large surfaces of the head; pain in one side of the head only: feeling as of waves of water, or something floating in the head; strong beating of the blood-vessels of the head; excessive sensibility to noise, light, and touch. For the following signs take *Aconite*:— violent, stupefying, compressive, and binding pain; burning pain through all the head: headache with buzzing in the ears, and running at the nose; headache made worse by speaking, drinking, or moving; made better by going into the open air; headache with feverishness, cold hands, and feet, and disturbance of the circulation. Use these medicines by putting five drops in a glass of water, and taking a teaspoonful every half hour, or hour, according to the severity of the pain. Observe the rules on diet. Read the next article.

Sick Headache.

Headache, with nausea, and vomiting, is a very common trouble, which many persons suffer with all their lives; when the cure is very easy. The remedy is Rush's Sarsaparilla and Iron. When an attack comes on, put a tablespoonful in a glass of water, and take a tablespoonful every half hour or hour; until relief ensues. To cure entirely:—take a teaspoonful unmixed, after each meal, and at bed time. *Common headaches* also may usually be permanently cured in this way. The head symptoms which indicate the use of Rush's Sarsaparilla and Iron are ;—dizziness; tenderness of the scalp; piercing and pressing pain on one side of the head; pain as of a nail driven into the head; periodical hammering and beating headache; pain in the back of the head; bursting open pain; falling out of the hair; heaviness of the head, with buzzing; *headache* with nausea and vomiting, or constipation. In very obstinate cases more than one bottle may be required, to effect an entire cure. The system will frequently be benefitted by taking the medicine, from three to six months, in all headaches, and neuralgia cases.

Offensive Breath.

This may be due to bad teeth. In that case have the bad ones removed; and those with cavities filled : and clean them daily with a good brush, and a little fine white soap. Every one for health, should take good care of the teeth. This being done, if a bad breath persists, it is a sure sign of disordered stomach. The remedy is Rush's Sarsaparilla and Iron ; a teaspoonful three times a day.

Palpitations. Fainting Fits.

These troubles show a disordered circulation of the blood, and generally, a deficiency and poverty of it. At the time of their occurrence give aconite as directed for headache. But the remedy which

will infallibly work a radical cure, by renewing
the blood, and restoring the whole system, is Rush's
Sarsaparilla and Iron. It should be taken after
each meal. One bottle will generally effect a
cure.

Nightmare.

This is always owing to disordered stomach, or
impurity, and poverty of blood. It will invariably
be cured by Rush's Sarsaparilla and Iron, taken
as above ; observing the rules of diet laid down in
this book.

Palsy.

Confirmed cases of this affection are generally
given up by physicians. It consists in a loss of the
power of motion, or sensation, or both ; and is confined to one half of the body. In all such cases there
is more to hope for in the use of Rush's Sarsaparilla and Iron, than from any or all other medicines ;
many cases have been entirely cured ; and para-
lytics would enjoy such an improvement of health
from taking it, as would well repay them. Persevere with it, according to the directions with the
bottle.

Epilepsy. Falling Sickness. Fits.

A person seized with this disorder falls instantly
wherever he may be, and is unconscious, and
convulsed ; with set teeth, frothing at the mouth,
clenched hands, and staring eyes. The neck tie,
and waistband should be loosened, and care taken
that the patient does not injure himself by his

struggles. A fit only lasts, in general, for a few minutes; it then passes off of itself. If such attacks occur frequently, *the mind* will be seriously injured. The cure may be set down as *sure* from using Rush's Sarsaparilla and Iron ; it should be persevered with, taking care to observe the directions, as to Diet laid down in this book. Recovery will be assisted by the morning Sponge Bath.

Drunkeness. Delirium Tremens.

The latter is caused by the former. It is usually attended by a very different kind of delirium, or craziness, from that of Brain Fever. *The tremens*, is a *busy*, helpless, talkative, timid, and cowardly delirium. The patient continually sees devils, imps, snakes, and the like. An attack may always be treated by giving two teaspoonfuls of *paregoric*, or fifteen drops of laudanum, once an hour, until relief is obtained. The patient should never be bled.

The Cure for Drunkeness.

The remedy is to add some nauseating drug to the liquor of the inebriate, without his knowledge. The best thing to use is Tartar Emetic, on account of its tasteless nature ; and because it produces prolonged, and severe nausea. As much may be added to a pint of whiskey, as can be held on the blade of a pen-knife, or about two grains : if that is found insufficient to produce nausea, and vomiting, the quantity may be doubled. After vomiting has commenced, a physician who is in the secret should be sent for, who should then use very strong moral suasion to induce a reformation ;—

saying, that when liquor produces such an effect, as nausea, and vomiting; a continuance of its use would probably prove fatal very soon. Many drunkards have been reformed in this way. A great disgust for rum is also frequently produced.

Common Cold. Catarrh.

This denotes a mild inflammation of the lining membrane of the nostrils, and windpipe, caused by exposure; especially with insufficient clothing. Children are particularly liable. The signs are slight fever, preceded by slight chills, or shiverings, with sneezing, obstruction or running of the nose, and eyes, some headache, and subsequently, hoarseness, rough and sore sensation in the throat, cough, wheezing, and difficulty of breathing. When this is confined to the nose, and frontal sinus, it is called *a cold in the head.* If the feet are cold they should be soaked in warm water (98 degrees,) until warmed. The patient should then go to bed and take a tablespoonful every hour, of a mixture made by putting a half a teaspoonful of tincture of Camphor in a teacupful of warm water. Place a hot brick at the feet, and encourage gentle perspiration by drinking a glass or two of cold water. This treatment will frequently break up an attack. Strong persons who have taken slight cold, may throw it off, by drinking one or two glasses of cold water on going to bed. If hoarseness, cough, and sore throat, seem to threaten an attack on the lungs, the remedy is Rush's Lung Balm; of which take a dose every two or three hours, until relief is obtained. If the cough continues, take the Balm three or four times a day. Many a *cold* has run on into *Consumption* for the want of a little proper medicine.

Influenza Grippe

This is only a *common cold* in a severe and epidemic form, attended with *great prostration of strength, and feeling of oppression* ; in addition to the signs of *common cold*, (see preceding article).

The treatment for this complaint is to be precisely the same as for *common cold*; except that in cases where there is much fever, Aconite (see page 151) should be given alternately with the Lung Balm. Put five drops of Aconite into a tumbler of water, and give a tablespoonful ; in one hour after put a tablespoonful of Rush's Lung Balm into another tumbler of water, and give a tablespoonful ; and so on, once an hour alternately, till the fever subsides ; and then give the Lung Balm alone every two or three hours. For a child give a teaspoonful of the mixtures, instead of a tablespoonful. If hoarseness, sore throat, and cough ensue, give the Lung Balm three or four times a day.

Quinsey. Sore Throat. Inflammation of the Tonsils.

The signs are heat, swelling, and redness of the back part of the throat, with difficulty of swallowing, impeded voice, and some fever. As the disease progresses the fever may increase, the tongue become foul, the tonsils very red and swollen, and white spots appear on them. The difficulty of swallowing is very great ; the tonsils often being so much swelled as nearly or quite to prevent swallowing anything. Soon after the appearance of the white spots on the tonsils, *matter* forms (denoted by a chill, or chills) and great relief is obtained as soon as it is discharged by lancing, or spontaneously breaking. The treatment is as follows:—if there is much fever, give Aconite and Belladonna (see page 151), alternately, by putting five drops of each in 1-3 tumbler of water (two glasses) and giving a teaspoonful of the aconite first, and after one hour, a teaspoonful of the Belladonna. If there is but little fever give the latter only. After one tonsil has broken, and discharged, the other may go through the same process. If the bowels are constipated open them with Rush's Pills. This disease shows a disordered stomach, and digestion, or impurity of the blood; which should be cured, after the tonsil has discharged, by taking a bottle of Rush's Sarsaparilla and Iron.

Diphtheria.

This disease resembles croup, with putrid sore throat. It is more frequent in changeable weather, than at other times, and arises from mainly the same causes with consumption. See causes of Consumption. Diphtheria should not be confounded with the putrid sore throat of Scarlet Fever; nor inflamed tonsils, or Quinsey. This disorder usually makes its appearance by a slight sore throat, rawness, and difficulty of swallowing. These symptoms rapidly grow worse, and there is a sensation as if of threads, or wool in the throat;—or of a bone which grows larger and larger every moment, and threatens suffocation. The face wears an anxious expression, and the usual signs of fever appear. The neck, and its glands swell, and swallowing becomes difficult, or is stopped entirely. In the mean time a false membrane, of a yellowish white color, which is characteristic of this disease,

12

forms rapidly, on the inside of the wind-pipe, and
extends down into the smaller air tubes, causing
the voice to become feeble, and shrill; and may
cause death by suffocation, in a period varying
from a few hours to several days. During the depo-
sition of this membrane, the tonsils, palate, and back
part of the mouth become inflamed, and more or
less covered with patches of white, or dead membrane. The back part of the nostrils also become
affected, and the breath very offensive. Great
prostration of strength attends the whole disorder.

Treatment. This should be as prompt as possible. Begin, in mild cases, with Rush's Lung Balm,
a dose according to directions, every two hours.
If the throat is hot and inflamed on looking inside,
it may first be packed, externally, with a towel
wet with nearly boiling water, applied as hot as it
can be borne : continue this every few minutes,
until the throat is very much reddened outside,
and then apply another, wet with cold water, extending down on the top of the chest, and up to
the ears, and cover it with dry flannel, pinned
closely around, this may be changed every half
hour, or hour, and kept wet, and cold, *when there is
fever*; but if not let the cloth become warm, and
sweat the throat.

Instead of using hot water to redden the skin,
volatile liniment may be used to rub the throat and
top of the chest freely ; and then be applied on a
flannel bandage round the throat. Warm the feet
by putting them in hot water, and keep them hot
with hot bricks, wrapped up in moist cloths. After
giving the Lung Balm for two or three doses, if no
improvement takes place, and there is fever, put
five drops of Aconite, (see page 151) in a tumbler
of water, and give a tablespoonful every hour ; if

there is any appearance of white spots in the throat, and a feeling of threads, wool, or a bone in the throat, with bad breath, dissolve a quarter of an ounce of Chlorate of Potash in a teacupful of water, and give a teaspoonful every three hours to an adult, and less to a child. Take this till free expectoration follows, or the throat is better. After that resume the Lung Balm every six hours. If the bowels are confined keep them open with Rush's Pills, at least for the first day or two.

In cases which seem malignant, or very bad, *with very foul tongue*, and bad breath, procure five grains of the yellow Iodide of Mercury; put it in a cup with a heaping teaspoonful of fine loaf sugar, and grind it up very fine with the end of a vial, or knife-handle; of this powder give as much as can be held on the blade of a pen-knife, every three hours, between the doses of Chlorate of Potash. If the Iodide of Mercury can not be had, use *blue mass* instead.

The throat inside should be swabbed with the following preparation; Chlorate of Potash 1-4 ounce; Muriate of Ammonia 1-4 ounce; mix with a half a teacupful of Syrup. Make a swab with cotton, or wool. the size of a peanut, tied on the end of a stick, or quill, and apply this mixture thoroughly, as far down as possible, every half hour, hour, or two hours; at the same time giving the Chlorate of Potash, and Iodide of Mercury internally. As the patient begins to be convalescent, he should return to the Lung Balm, to prevent any injury to the lungs. The diet should be very light and simple as for fevers; but in long continued cases, where there is great debility, beef tea, and thin chicken soup may be cautiously given. Such cases will be most rapidly restored to health by

Rush's Sarsaparilla and Iron, as soon as the throat disease is subdued.

The above treatment should be followed energetically, even in very bad cases; for there are very few indeed, we might almost say none at all, which are not curable, as above.

Pains, or Spasms in the Stomach.

These are known by contractive, spasmodic, or gnawing pains at the pit of the stomach; attended with anxiety, nausea, belching, or vomiting; with faintness, and coldness of the hands and feet. Various other dyspeptic signs are often present; (see Dyspepsia,) and the causes are mainly the same. This complaint is more common in females, than in males; and is due to any interruption, or disorder, of the usual monthly flow. Fortunately we now have a very safe, and agreeable remedy, for this destressing complaint, in Rush's Sarsaparilla and Iron. It should be taken after each meal; and in case of an attack of pain, put two table-spoonfuls in a tumbler of water and take a table-spoonful every half-hour. The morning sponge bath (see page 14,) will much assist the cure. Also observe the rules of diet, &c., laid down in this book.

Heartburn, Water-brash,

Is a painful or uneasy sensation, of *heat* or *burning*, at the pit of the stomach, sometimes extending upwards. It is accompanied by the symptoms described in the last article, and relief is obtained on throwing off a small quantity of limpid liquid. The remedy is Rush's Sarsaparilla and Iron, which should be taken as just directed. A permanent cure may be expected.

Vomiting of Blood.

Blood evacuated in this way is usually of a dark color, or nearly black, and in clots; or may be mixed with bile, or food. It is sometimes discharged by stool. This loss of blood may be caused by an aggravated form of dyspepsia; and is then preceded by many of the sings of that disease. It sometimes follows a suppression of the female flow, or stoppage of bleeding piles. The blood, in cases of vomiting of it, usually *oozes* through the coats of the bloodvessels of the stomach: and is *not* lost from a rupture of the bloodvessel, as is usually supposed. In case of an attack, give an even teaspoonful of fine salt, dissolved in a little cold water. This will check, or stop the bleeding; and as soon as that is done put two tablespoonfuls of Rush's Sarsaparilla and Iron into a tumbler of water, and give a tablespoonful every three hours. After the attack is over, continue this medicine, as directed on the bottle. If persevered with, it will restore the system, by purifying the vital fluids, and thus prevent another attack.

Consumption of the Blood. Anæmia. Leanness.

Any exhausting, and debilitating causes may produce emaciation, debility, leanness, and poverty, and thinness of the blood. The blood becomes more watery than in a healthy state, and less rich in the red globules, or blood disks, or coloring matter. (See page 48.) Physiology has conclusively shown that a certain proportion of Iron, in an exceedingly delicate, and soluble condition, is an *indispensable* item, in the red globules, or blood disks, which are the very life of the blood, and through that vital fluid, of the whole system. Without this proportion of Iron, from failure to obtain it in sufficient quantity in the food, or from want of power in the digestive organs, to take it up, the blood no longer renews, and reproduces its red globules and disks, and they, in turn, are unable to restore the daily waste, and wear and tear of the whole body. The best remedy ever discovered is Rush's Sarsaparilla and Iron. The Iron in this medicine is in the very best state of chemical combination to form blood disks ; while the Sarsaparilla drives out all corrupt humors, and strengthens and invigorates the stomach, and digestive organs; enabling them to supply well digested food to assist in this wonderful and beautiful process. In many cases one bottle will prove sufficient ; but in cases of great emaciation, and debility, arising from long fever, loss of blood, wounds, self-abuse, Scrofula, secondary Syphilis, premature decay, and the like ; a course of from three, to six months, or even a year, may be necessary for a perfect cure. In such cases as the above, the soft glow of returning health, will richly repay patience, and perseverance.

Dropsy.

By this we mean a collection of serous, or watery fluid, under the skin, or in any of the cavities of the body. When under the skin we know it by the swelled, bloated and waxy appearance ; and by the finger making a *pit*, when pressed firmly on the skin, which afterwards disappears slowly. A dropsy of the internal cavities, as the head, chest, and abdomen, is much more dangerous. All dropsies are more or less due to Scrofula ; (page 40.) or a Scrofulous condition of the blood and system. The remedy is to purify the blood, and rid it of this source of complaint. Dropsy is always accompanied by a poor, and watery condition of the blood, akin to Anæmia, or *Consumption of the Blood*, (page 182.) they both being caused by Scrofulous, and Syphilitic taints ; either acquired, or hereditary. All cases of Dropsy may be partially, or entirely cured by thorough purging, and by Rush's Sarsaparilla and Iron. Take four of Rush's Pills every second night for three doses ; or enough to move the bowels three or four times, for a dose. At the same time take Rush's Sarsaparilla and Iron, after each meal, and continue to keep the bowels open by an occasional dose of pills. All curable cases can be cured in this way.

Cancer.

There are two or more kinds of cancer. The most common is the *hard kind*, which begins with a hard knotty tumor, or swelling, and gradually increases in size, with shooting, and darting pains. By the time these tumors have reached the size of an orange, they are exceedingly painful, allowing little rest by day, or night. In this condition they are called *rose* cancers. This is all a slow process ; it being usually three or four years from the beginning, before they prove fatal. · There is, however, no necessity for this result ; since a hard cancer can always be removed, by Rush's

Cancer Plaster, before it has become an open sore,
and sometimes afterwards. The cause of cancer
seems to be a bad habit of body, similar to Scrofula, and Syphilis ; indeed some eminent medical
men are of opinion that cancer is a combination
of both, either transmitted hereditarily, or not.
This cancerous habit of body may be known by
a sallow, leaden, and waxy hue of the skin ; and
by emaciation, debility, bad appetite, and imperfect digestion. In such a person, when a very
hard tumor appears, with fits of darting pain ;
cancer may be supposed. It may be checked,
or kept quiet for years as follows :—take a quart
measure crowded full of the little plant Pyrola or
Pipsissewa, gathered roots and all ; put it all into
three pints of soft water, and boil to one pint and
strain. Mix this with one bottle of Rush's Sarsaparilla and Iron, and take a tablespoonful after
each meal. This may be continued as long as the·
health improves. Much benefit nay be derived
from this, even after the cancer has become an
open sore. An open cancer should have very simple dressing, such as simple cerate spread on linen,
or compresses wet with tepid water. Persons interested should also read the articles on· Scrofula and
Syphilis. Those wishing any further advice, or
information, on this subject, may recieve it gratu-
itously by addressing the editor ; see page 44.
Sufferers are particularly cautioned not to fall into the hands of so called *Indian Doctors*, who are
ignorant, and mercenary pretenders, almost to a
man.

Canker in the Mouth.

In this affection the mouth has a bad smell, the gums are hot, red, swollen, and spongy; with a viscid, or bloody discharge; or ulcerated next to the teeth, which are perhaps loose; there may be general debility or slow fever. This disease is of the nature of scurvy, and is caused by improper diet in the use of salt food, or a lack of vegetables. The blood is always deficient in iron. A cure may be easily effected by taking Rush's Sarsaparilla and Iron. At the same time the gums may be rubbed with tincture of myrrh, three or four times a day.

White Canker spots in the mouth should be touched lightly with a small pointed piece of nitrate of silver; when a cure will immediately result. But Rush's Sarsaparilla and Iron should be taken also; to prevent a return; and to remedy the accompanying scorbutic tendency of the system, and restore the blood.

Scurvy.

This is characterized by great debility, pale and bloated face, swelling of the feet and legs, bleedings, livid spots and ulcers of the skin, oppressive urine and stool. The gums are spongy as described in the preceding article. It is caused by exposure, and improper diet; also by intemperance, impure air, and unclean liness. To cure such cases a wholesome diet must be adopted, with vegetables and acid fruits; while Rush's Sarsaparilla and Iron will renovate the blood, and restore the entire system.

Earache.

This is a very common complaint with children, and frequently attended with very severe pain. It is usually a sign of Scrofula; (see page 40). A very good remedy is a poultice made of a roasted onion, slightly mashed, and applied very warm, with twenty drops of Laudanum on the surface. Where the external ear is much affected put five drops of Pulsatilla in a half a tumbler of water, and give a teaspoonful every half-hour. This, with the poultice, will rarely fail to give relief. The Scrofulous tendency of the system, upon which it depends should be removed by taking a bottle of Rush's Sarsaparilla and Iron.

Inflammation of the Eyes. Sore Eyes.

The whites of the eyes are red and inflamed, as well as the lining of the lids; and there is a feeling as of sand in the eye. A discharge also takes place, and the eyes are stuck together, especially in the morning. To relieve this annoying trouble take equal parts of Elderblows, Chamomile flowers, and Rose leaves, a handful of each, and having steeped them, fifteen minutes, in hot water enough to cover them, let the whole get cold, and then put the whole a little squeezed out, between two layers of thin muslin, and bind it on the eyes as a poultice. As often as it gets hot or dry, wet it with the liquid, squeezed out, or cold water. This may be worn only at night; or day, and night, if the eyes are bad enough to require it. This application is far better than any eye waters, or washes. Physicians are recommended to try it. *Eye complaints* show a Scrofulous tendency in the system, which should be promptly eradicated by Rush's Sarsaparilla and Iron.

Erysipelas. St. Anthony's Fire.

This is a disease of the skin, and tissues which are directly under the skin; the signs of which are heat, tingling or pricking pains, with extensive swelling, tightness of skin, and deep-colored, shining redness. As the disease progresses the pain becomes pungent, burning, tearing and shooting, and is made worse by motion, and pressure. There is generally feverishness, foul tongue, nausea, headache, oppression at the stomach, and sleepiness. In a few days blisters filled with clear watery serum appear on the skin. *Erysipelas, of the face and head,* is considered the most genuine form; and usually begins with the nose, or an ear, which appears dark red, and much swelled. From this point it rapidly spreads; delirium may come on; and the swelling is often so great as to close up the eyes, and completely disguise, and disfigure the face. This disease is not without danger; and is thought to be contagious when the breath of the patient is directly breathed. It will terminate quickly, and favorably, if treated as follows;—as soon as the pain, redness, swelling, heat, and headache commence, put five drops of Belladonna (see page 151), in a glass of water, and give a tablespoonful every hour. If the swelling increases; and particularly if blisters appear, put five drops of Rhus tox (see page 151) in a tumbler of water, and give a tablespoonful, and in one one hour after the Belladonna, and so on, once an hour, alternately. The worst attacks will usually be cured in this way, in less than a week. The feet must be kept warm; and the burning pain may be relieved by linen cloths wet with cold water, and laid on the parts; or, apply a poultice made of stewed cranberries and crackers, to be kept moist. If no Rhus Tox is at hand, during the summer season it may be prepared as follows:—put five leaves of the *poison ivy,* the low plant, (not the creeping kind) in a half pint of boiling water, and steep half an hour, of this put a tablespoonful in a tumbler of water, and give as ordered. It is a safe, and powerful remedy, against Erysipelas.

The bowels should be opened at first, by Rush's Pills. After an attack has subsided, the Erysipelas humor may be eradicated from the system, by Rush's Sarsaparilla and Iron; and a second attack prevented; which would otherwise be very likely to succeed, after a longer or shorter period.

There are various eruptions called *Erysipelas humor*, not attended with swelling, but with redness, itching, burning, and smarting, of which the different kinds of *Eczema* are the chief. These all indicate a humory condition of the blood, which if neglected may run into Scrofula; or break out in confirmed, and dangerous Erysipelas of the face, and head. Do not neglect to eradicate all such humors from the system by a sufficient course of Rush's Sarsaparilla and Iron. See page 47.

Asthma. Phthisic.

This is a very distressing disorder affecting the breathing. It is caused by spasm of the very small air tubes, and cells, by which the air is kept from entering the air cells of the lungs. Attacks more frequently come on in the night; and last from an hour, or two, to a day or more. Most attacks end with expectoration of mucus, and are called *moist* or *humid*; others remain *dry*, and are so called. Such attacks are liable to increase in frequency, and sometimes torment the patient nearly every night; preventing lying down almost entirely. Cases have come within our knowledge where for months, and years, the sufferer could not lie down; *one*, a lady was cured by Rush's Lung Balm, and other medicines in about five weeks. All cases under the new treatment are curable; but the remedies should be adapted to each individual case. Some are cured much sooner than others; and no satisfactory directions can be given in a work like this. Out of 157 cases, on our note book, it appears that all were cured but three, who are now rapidly recovering. Persons wishing to avail themselves of the same treatment, may find the editor's address at page 44.

Diseases of the Heart.

These are usually considered fatal, by the medical faculty; but our own experience of their treatment proves their curability. We could easily cite many cases in support of this assertion. Heart disease is characterised by many strange, and uneasy feelings about the left side of the chest; with irregular action of the heart. The pulse shows an intermission between the beats; and on listening over the heart, various odd sounds may, in some cases, but not all, be heard. Patients compare the sensations to beating, fluttering, throbbing, panting, or thrilling. Palpitations are very common; and all these signs are apt to be made worse, by going up stairs, any sudden exertion, or heavy meals. Sudden sharp pain darts through the chest, and region of the heart; nervous attacks are common; and sudden starting on going to sleep. Scrofula, and dyspepsia aggravate such troubles. We again repeat that the great majority of cases are curable; but that the treatment must be adapted to each case. Persons wishing to avail themselves of the experience of the editor, can consult him free of charge; address at page 44.

Eczema. Salt Rheum. Humid Tetter.

This is an eruption characterized by blisters, small and in patches, containing clear watery, or slightly opaque matter; and by chafing, or excoriation, giving rise to more or less running, which hardens into scabs; terminating by desquamation of the scarf skin. The skin looks red, angry, and inflamed; and there is much burning, and burning itching. There is usually no swelling. This eruption may appear in almost any part of the body, and sometimes extends over the whole person. When it appears on the face, or head, with much running and scabbing, it is called *Impetigo*. Such cases require a course of purging by giving Rush's Pills every night, or every second night; dose, one or two pills, or enough to procure three, or four stools a day. At the same time do not fail to take Rush's Sarsaparilla and Iron. These medicines should be continued until improvement begins, when the pills may be left off; but the Sarsaparilla and Iron should be continued until a *complete cure* takes place. *Do not make any greasy applications*; but use washes of soft water with a little thin starch in it. You may boil about a teaspoonful of dry starch in a quart or two of water, and use it cold as a wash; or applied on cloths, to allay the burning and itching.

Lichen. Dry Itch.

We know this by an eruption of red pimples ranged in rows, attended by redness of the skin which soon disappears, leaving it *thickened, rough, and with its folds and creases enlarged.* A slight dark red color may remain, and there is usually *severe* itching, especially at night. It usually prefers the hands, fore-arms, neck, and thighs for its seat; but may take any part of the body. This eruption is not caused by an *itch-insect*; and can not be cured by greasy ointments. The very best mode of treatment is to proceed just as recommended for *Eczema*, in the preceding article. The itching may be allayed by the wet cloths; and the disease removed from the blood by Rush's Sarsaparilla and Iron.

Psoriasis.

Is a disease of the skin characterized by silvery, white scabs folded over and over; these are very adherent to the skin, and cover a dark red, thickened and slightly projecting surface. It may be in separate patches of large or small extent; or it may cover large surfaces. When present at all, *it is found on or about the elbows or knees.*

This affection shows a morbid condition of the blood, which can be removed, and the disease cured, by Rush's Sarsaparilla and Iron. The length of time will depend on the inveteracy of the disease. Persevere in its use, and you will be well repaid by returning health.

The Itch. Psora.

This is caused by a little animal, too small to be seen, except by the microscope, which burrows in the skin. An eruption of small vesicles, or blisters, attended by severe itching, especially at night, when warm in bed.

It is communicated from one person to another by a transfer of one or more of the little animals. These usually prefer the spaces between the fingers, bends of the elbows, the nipples of females, or penis of males; and other parts where the skin is thinest. It is easy to get rid of; since we have in Sulphur a sure remedy. Proceed as follows; procure one ounce of washed sulphur, and melt it in four ounces of lard, stirring it in until well mixed. When nearly cold, half a teaspoonful of oil of bergamot, or any convenient perfume may or may not be added. If you can not get washed sulphur, take the common flowers of sulphur, and wash them, by putting them in a basin or bowl of soft water, letting the sulphur settle, pouring off the water, and then drying the sulphur, on coarse paper.

To cure the itch, rub a little of the ointment over the affected places, on going to bed at night. In the morning wash it off with soap, and soft water. Repeat this for three nights, and then change all the linen. If all the little animals are not killed; it may be necessary to repeat this process. This is better than ointing the whole body at once. The above quantity of ointment is enough for a large family. The presence of itch is usually accompanied by more or less disorder of stomach, and digestion, as well as poverty of the blood;—in all such cases, it will be well to strengthen, and invigorate the system, by taking a bottle of Rush's Sarsaparilla and Iron. See page 47.

Shingles. Zona.

This is a disease which makes its appearance on the skin in clusters of small vesicles, or blisters, on bright red patches of skin, extending half round the body, like a girdle, perhaps as wide as the hand. The right side of the body is more commonly affected, than the left. There is considerable pricking, smarting, and even sharp pain. There is generally some fever, headache, nausea, and loss of appetite.

The vesicles, as above described, grow larger, perhaps to the size of a pea; break, discharge, and dry up; the red color of the skin remaining some time after; and the entire disease continuing three or four weeks. This complaint requires no serious medication; but in case of all such eruptions there is more or less impurity of blood, which will be renovated by the use of Rush's Sarsaparilla and Iron.

Boils, or Biles. Carbuncle.

It is not necessary to describe a boil, of which a carbuncle is only a large one. When one begins it should be opened, the earlier the better, even before matter has formed. This will let it bleed, and diminish the size of it. Then apply a towel or compress, wet with cold water, having a few drops of tincture of Arnica in it, (page 151,) and put a dry one on outside. Keep the inside one wet. This is better than poulticing, and will resolve, and bring the boil to a head, much sooner. *A Carbuncle* is an immense boil, which seems to prefer the *back* for its seat; but may come elsewhere. It should be treated as just recommended, except that a poultice of stewed cranberry, and pounded charcoal, should be used instead of wet compresses, if there seems a tendency to black spots, or mortification. In both these affections there is a great impurity of blood; which will require a course of Rush's Sarsaparilla and Iron to eradicate.

Abscess.

This is a collection of matter which forms in glandular and muscular parts of the body; as in the glands of the arm-pit, or female breast; in the muscles of the thigh, &c. They may break of themselves; but it is always better to have them opened; as this makes them smaller, and there is less matter formed. Compresses wet with tepid water, and kept wet, and covered with a dry one, are the best application for resolving an abscess, or bringing it to a head quickly. A few drops of Arnica may be put in the water, with advantage. The formation of matter is usually attended by one or more chills. An abscess always shows an impure condition of the blood; which, if not eradicated, may result in disease of the lungs, or some other internal organ. This purification may

13

always be effected by a course of Rush's Sarsaparilla and Iron.   One bottle will usually suffice.

Felon. Whitlow.

This is an abscess forming in the end of the finger. It should be opened early, and may be poulticed with cranberry, or with wet linen.   They are very painful, and show a corrupt state of the blood, requiring the same treatment as laid down in the last article; which see.

Corns.

Should be pared close with a sharp knife, and have Arnica plaster applied, see page 151.   Be careful to wear large boots.   Those made on *Plummer's* last, are easiest.   To soft corns, apply Arnica tincture, or plaster: page 151.

Chilblains.

These are caused by cold.   The irritation may be relieved by applying tincture of Capsicum, (hot drops); or by soaking them in hot water, in which a large bunch of sweet fern has been boiled.   Two applications of the latter will usually cure them for the season   .

Ulcers.

An ulcer is a sore.   It may have a variety of causes; but generally shows an impure or impoverished state of the blood.   The best dressing for an ulcer is some soft linen, wet with soft water, containing a few drops of Arnica, or walnut leaves.   Keep it constantly wet, and a dry cloth on outside.   The healing of ulcers will be very much promoted by purifying the blood, and improving the tone of the stomach and digestive organs.   This will be best effected by Rush's Sarsaparilla and Iron.

DISEASES OF CHILDREN AND INFANTS.

Scarlet Fever. Scarlatina. Canker Rash.

This is a highly infectious and contagious disease, characterized by high fever, and the appearance over large surfaces, or the whole of the body, of an eruption of a bright red color, *smooth and glossy*, which appears on the second or third day of the fever. It is frequently complicated with ulcerated, or putrid sore throat, resembling diphtheria, which see. The fever is frequently *violent*; but begins to subside in about five days; when the eruption also fades away, and ends by the cuticle, or scarf skin peeling off, in large or small pieces. In the most severe cases the eruption is *slow* in appearing and scanty; and the inward organs are severely affected. The head and throat, are most affected; the putrid character of the disease seeming to extend from the throat to the brain. Most cases of Scarlet Fever are cured very promptly by Belladonna. (See page 151.) Put five drops in a glass of water, and give a teaspoonful every hour, and so continue, until the disease subsides. *When this disease is prevailing*, give the same medicine every six hours, to the well ones as a *preventive*; and to ernder an attack *milder* if it should occur. The whole family of children should take it; and very many cases of Diphtheria will be prevented, and lives saved. Adults who have not previously had this fever, should also take Belladonna in the same way.

Care should be particularly taken not to allow children to be exposed to taking cold, *while recovering*; as such exposure frequently brings on a sort of dropsy, which proves fatal. Children who are left weak, and sickly, will be rapidly restored to health, by the use of Rush's Sarsaparilla and Iron.

Measles.

This is also a contagious eruptive fever, the eruption coming out on the third, fourth, or fifth day;
the color of which is a *dark* red, and is generally seen
in *lentil shaped spots.* The fever is usually much less
severe than in Scarlet Fever. It is considered a
more severe disease for adults, than for children. A
short dry cough, sneezing, and redness of the eyes,
generally precede the eruption.

Most cases are promptly cured by the following
treatment ;—put five drops of Aconite (see page 151)
in a tumbler of water, and two teaspoonfuls of Rush's
Balm in another tumblerful; give a teaspoonful of
Aconite first, and in one hour after, a teaspoonful of
Balm mixture, and so on. Adults may take a tablespoonful. Give only once in two or three hours, as
soon as improvement takes place. When the eruption is slow in appearing, especially in adults, or when
it seems to be imperfect, or to have receded, or struck
in, the patient should take a sweat by means of the
vapor bath; or in any convenient way. It should
be followed by sponging the skin all over with tepid
water. A sweat will invariably bring out the eruption, and with Rush's Lung Balm, render the disease
safe and mild. The Balm will also prevent it from
settling on the lungs, in persons of consumptive family. Persons left in a weakly, and sickly state, by
measles, will recover rapidly by taking Rush's Sarsaparilla and Iron.

Small Pox. Varioloid. Vaccination,

Small Pox is also an eruptive fever, which comes
on in twelve or fourteen days from the time of exposure to contagion, the eruption appearing on the fourth
day of the fever, which begins with the usual signs of

fever ; besides much *backache*, and pain at the pit of the stomach. The eruption begins on the face, forehead, and in the edge of the hair, as red pimples, or pustules, with a depression in the centre, as they enlarge, filled with semi-transparent fluid. When fully filled, they become as large as small peas. The violence of the disease depends upon the number of these pustules. In the worst, and fatal cases, they are so numerous as to run together. We may distinguish this disease from any other, by a small lump like a grain of wheat, to be felt, the first day in the centre of each of the pimples, or pocks. About three days are taken up by the stage of fever ; five days by the pocks forming, and filling with matter ; four days by scabs forming on them ; and four days more by the scabs falling off.

The most dangerous time is the period of scabbing. *Varioloid* is Small Pox, modified by previous vaccination. Most cases of Small Pox will be cured by Aconite, and Lung Balm, precisely as directed for measles, which see. In addition, procure, if possible, a dozen fresh leaves of the *Pitcher Plant*, or half an ounce of the dried plant, and steep them in a quart of water, and give two tablespoonfuls every three hours. This, however, will not be needed except in very bad cases. The Pitcher plant can usually be obtained at Botanic Drug Stores. To prevent pitting, keep the face constantly covered by linen cloths, wet with tepid water : this will succeed better than gold leaf, or other means. The room should be *thoroughly* ventilated, and bed changed daily. A plenty of cold water should be allowed the patient to drink. After the pustules are fully formed, omit the Aconite, and give the Lung Balm, and Pitcher Plant, two hours apart. Varioloid has the pustules generally a little smaller, and much fewer in number. The treatment should be mild, but precisely the same. *Vaccination* was discovered by Dr. Jenner, in 1796. It is the same disease passed through the cow. It is performed by placing, with the point of a lancet, or needle, a little of the clear serum, or matter, from the cow, or a child, (kine pox) under the scarf-skin. A very small piece of the dried scab will do. In one week the kine pox will be complete; and the pustule resembles very much a small pox pimple, and scabs in the same way. Vaccination generally protects for life : but should be tried again after every alarm of Small Pox : since in many cases it fails, after a longer, or shorter time. The time for obtaining the clear matter, for vaccinating, is on the 7th, 8th, and 9th days of the vaccination.

Chicken Pox.

This is also an eruptive fever, the eruption of which has some resemblance to Small Pox. The fever, however, is mild, and the pustules are ripe by the third day, and the whole eruption disappears by the end of the fifth, without leaving any marks. It does not usually require much treatment. If there is much fever, give Aconite; if the head seems affected, Belladonna; (see page 151,) if there is cough, wheezing, or oppression of the lungs, Rush's Lung Balm. In common cases put two teaspoonfuls in a glass of water, and give a teaspoonful every hour.

Nettle Rash.

This is an eruption of whitish spots like nettle stings, appearing and disappearing suddenly, by heat and cold ; and often changing place. These are usually signs of Dyspepsia, which see ; and this disorder is generally caused by a Scrofulous and humory condition of the blood ; and requires for its eradication a regular course of treatment. The very best is Rush's Sarsaparilla and Iron.

Hooping Cough.

This hardly needs much description. It consists of violent, and convulsive, expirations, interrupted by long whistling inspirations, and ending in a shrill whoop; followed by expectoration of mucus, or a fit of vomiting. In severe cases suffocation is threatened, and there may be bleeding from the nose, mouth, or ears. Such fits of coughing return, perhaps, every three or four hours, or oftener. Hooping Cough is promptly cured, or rendered very mild by Rush's Lung Balm. Put two teaspoonfuls in a glass of water, and give to a young child a teaspoonful every two hours ; or for an older child two teaspoonfuls ; or it may be given by drops, according to directions with bottle, *after each coughing spell.* If there is vomiting, danger of suffocation, and accumulation of phlegm ; put five drops of Ipecac (page 151,) in a glass of water, and give a teaspoonful alternately every hour, or two, with Rush's Lung Balm ; that is, the Balm one hour, and the Ipecac the next. As soon as relief ensues, leave off the Ipecac, but continue the Balm. Put the child on a strict diet, of light, but nutritious, and unstimulating food. See page 22. Many cases of this Cough leave the little sufferers very much reduced in flesh and strength ; all such may be very soon restored by Rush's Sarsaparilla and Iron.

Croup.

This is an affection of the throat, in children ; which we know by short, difficult, and hoarse breathing : with a shrill, harsh, metallic cough, like cough-

ing through a brass trumpet; the little sufferer
throws his head back; there is sometimes fever, and
the brain affected. *True* or *membranous Croup* consists of an inflammation of the lining of the trachea,
or windpipe, causing the secretion of a thick, viscid
substance, opaque, and like the boiled white of egg.
This adheres to the inside of the windpipe, takes the
same form, and is called a *false membrane.* The
same thing takes place, but more extensively, and
higher up, in true *Diphtheria*; (which see.) In
croup, when a false membrane has been allowed to
form the case is extremely critical. Exposure to
cold, and improper food, and stimulating diet, seem
to be the causes. The signs of croup are at first
those of a common cold, with some fever; but in a
day or two the cough begins to take on the character already described; and sometimes resembles the
crowing of a young cock. Fever, with the peculiar
cough, shows that inflammation has begun; and that
no time should be lost. After this a variable state of
fever, and restlessness continue; the face looks very
anxious, and changes from red to a livid hue; and
the fits of coughing are followed by a profuse and
clammy perspiration. As danger increases, the
breathing becomes more and more difficult, the
cough assumes a more husky tone; the voice sinks
to a whisper; the eye looks dull, and glassy; and
there is great prostration. *True* croup generally
comes on in this way, *slowly* and *insidiously*; and
such cases are the most dangerous. *False* or *spasmodic* croup comes on suddenly, and seems more
frightful than the real or *membranous* form. In false
croup there is no *false membrane* formed; but the
symptoms of cough, and difficult breathing are much
the same. Such an attack frequently comes on in
the night, without any warning; and a child may be
subject to such attacks. The *treatment* should be as

follows :—as soon as any such signs of cold, or cough are observed, give Rush's Lung Balm, in doses according to age, every two or three hours, a few doses will generally break up an attack ; if it should not, give enough *Hive Syrup*, or wine of Ipecac to vomit.   A young child may take a teaspoonful, an older one two teaspoonfuls, for a dose ; to be repeated every ten minutes until vomiting ensues.   After that, go on with the Lung Balm.   If there should be much fever, put five drops Aconite (page 151,) in a glass of water, and give a teaspoonful ; then in one hour give the Lung Balm, and so on until the fever abates. This course will rarely if ever fail, to break up the attack.   A *sudden attack* of croup, may be broken up by doses of Lung Balm once an hour ; and by putting a towel wet with cold water round the throat ; or by vomiting with wine of Ipecac, or Hive Syrup. In such a case give the Lung Balm, the next day, every three hours, to prevent another attack.   In using the wet towel round the throat, which is an excellent remedy, keep it wet and cool, and a dry one outside.   Never neglect this remedy, when there is fever, and signs of heat about the throat.   When cases have got into the worst condition described above, Bromine is the remedy.   To prepare this put five drops of Bromine in an ounce vial of water, and shake it well.   Put five drops of this in a glass of water, and give a teaspoonful every half-hour.   This is a Homœopathic remedy, and has succeeded in very desperate cases.   Children subject to attacks of croup, are *humory*, or *scrofulous*, and should have this eradicated by taking Rush's Sarsaparilla and Iron.

Rickets.

This malady usually begins at the age of one or two years, and is known by increasing deformity of the osseous, or bony system. The head is usually large ; the breast bone, and forehead prominent ; ribs flattened, and the long bones, and spine distorted. The digestion, and bowels are disordered, and the muscles become wasted, and limbs emaciated. If the disease is not arrested, a fatal termination may result. The remedy is Rush's Sarsaparilla and Iron, which should be given according to age, and the directions accompanying each package.

Dropsy of the Brain. Hydrocephalus.

This malady is usually found in Scrofulous children, and is in fact caused by *tubercles* forming at the base of the brain. It is, for the child's brain, what *consumption* is for the lungs of adults. Such children usually have very large heads ; and would probably fall victims to some other form of scrofula, if not to this. An attack generally comes on in a slow and insidious manner ; and the slight feverishness, and peevishness is assigned to some such cause as teething, or disordered stomach. Sometimes, however, the symptoms are more striking, and resemble simple Inflammation of the Brain, (which see.) In general there is fever, especially at night ; the child becomes peevish, *when raised up from lying down* ; it may have fits of screaming, grinding the teeth, redness of the face, and eyes, and convulsions, and stupor. This is an invariably fatal disease, when fully formed ; but if treated early, by the most powerful anti scrofulous remedies, a different result may be hoped for. The best remedy for this purpose, as well as for warding off, and eradicating every form of

Scrofula, in its early stages, is Rush's Sarsaparilla and Iron (see page 47). It is peculiarly adapted to the organism of females, and children. The mother should usually take the same medicine.

Convulsions, Fits.

This is peculiarly a disease of childhood, and arises from the physical peculiarities of infancy, in the preponderance of the nervous and central system. The immediate causes are scrofulous humors, worms, colic, disordered stomach, teething, and all irritating causes. A child in a fit, should have the feet, and legs, placed in water at about 110 degrees; or nearly as warm as the hand can bear; and at the same time sop the top and back of the head very freely with cold water. Continue this for eight or ten minutes, or until the fit seems somewhat, or entirely subdued; then place the child in a dry warm wrapper. This process should be *repeated again, and again*, in case of long fits, until the child comes out of it. When fits seem to be caused by improper food, the bowels should be cleared by large injections of warm water, given at the same time with the bath. The best *medicine* for most fits of children is *chamomile*; put five drops (see page 151) in a glass of water, and give a teaspoonful every fifteen minutes, until the child comes out. If the eyes, or head seem affected, give Belladonna, in the same way, instead, or after the other. These medicines may be given at the same time with the bath; but after the child is better, only once in three or four hours. If no other medicine is at hand, give Camphor in the same way; *instead*, but not *with* the other medicines. *Convulsions* uniformly show a tendency to Scrofulous humors, or their actual presence in the system. Hence the absolute necessity of purifying the blood, in order to avoid more fits, and other diseases. *The remedy* is Rush's Sarsaparilla and Iron. If the child is nursing, the mother may take the medicine, or both mother and child. The mother's health will generally be as much improved as the child's.

Care of New Born Infants.

A new born infant should be gently washed with soft water, warmed to blood heat, or 98 degrees. No soap should be used; but a little bran is allowable, and some lard, to assist in removing the sticky substance, of a whitish yellow color, which covers more or less, their little bodies.

The child should have a thickness of fine linen, or cotton, next the skin, and soft flannel outside. The clothing should be warmed, and all changes of temperature, or any chill should be carefully avoided, during the first few days, or until the child has become accustomed to such a great change of temperature, as takes place at birth. The infant should continue to have a daily bath in *soft* water, lowering the temperature one degree every two days, beginning at 95, and going down to 75. The whole body should be immersed; and three minutes is long enough for the bath. All the clothing about a baby should be loose; and all swathing and bandaging *absolutely prohibited*. The quantity of clothing must be abundant at *first*; but should be gradually diminished after the first few days.

The baby should be put to the breast in six hours after birth, as this will assist in moving its bowels, even if no milk is obtained. *This is also beneficial to the mother.* If it is thought desirable to give the babe any nourishment, before milk comes, take one part cream, or rich milk, and three parts warm water, and sweeten *a little only*, with molasses. This should be given, at first, at intervals of three hours. After the babe's bowels are moved, substitute a very little loaf sugar for molasses. The proper time for weaning is nine months after birth; as the milk then becomes less nutritious; and all nursing after one year old is very injurious, to both mother and child. Many mothers continue nursing for no other reason than to prevent another pregnancy. To all such we would say, do not spoil your health and beauty, and bring on whites, falling of the womb, and consumption of the blood, by this pernicious practice, but instead, use Rush's Restorer and Preventive, (page 231), and *wean your baby.* If the mother does not have milk enough, the babe must be supplied from a bottle, kept very sweet and clean, with rich cow's milk, and water sweetened as previously directed. About once in four hours will be often enough for nursing, or feeding. The process of weaning should be gradual, and farinaceous food may be given, in small but increasing quantity; such as rice, sago, tapioca, arrow root, and by degrees ordinary unstimulating diet.

Sleep. Want of Sleep. Wakefulness.

The babe should sleep by its mother's side for the first six weeks, in order to keep it uniformly warm. But after that period a separate cradle or bed will be better. It is not healthy for children to sleep with old people, for any length of time. After the first few days the babe's head should not be buried up when sleeping, but a plenty of fresh air allowed, by leaving the nose and mouth, at least, free. There is usually an unreasonable fear of *taking cold in the head.* A baby should be rocked but little, if at all; as rocking disposes to fits, or head disease. Paregoric, soothing syrup, and opiates generally, should be wholly banished, as likely to produce permanent injury to the brain, and whole system, if persisted in. A nurse employing anything of the kind, should be immediately dismissed. And no parent deserves the blessing of a child who will poison its little life with narcotics. In order to be better understood we will give the whole forbidden list. Paregoric, Laudanum, Syrup of Poppies, Godfrey's Cordial, and all preparations of Morphine. Do you say, what then shall I do for my crying, and sleepless child ? The answer is, *find out the cause, and remove it* ; but do not try to cover it up by opiates. For sleeplessness, where the nurse drinks no coffee, and no nurse should; put two drops of coffea (see page 151) in a tumbler of water, and give a teaspoonful every hour, until sleep is obtained. While weaning, use Belladonna in the same way. If there is diarrhœa, or during teething, use Chamomile (see page 151.) In many cases of restlessness both in adults, and children, a small dose of Rush's Sarsaparilla and Iron, will allay restlessness, and give healthy and refreshing sleep, when all other remedies fail. It contains no opium, or pernicious drug, and will always strengthen the system. It will never do harm to try it. Doses are stated with the bottle.

Exercise of Infants.

During the warm season babies should be carried much in arms, or a carriage, in the open air. In cold weather *crying* seems to be their natural exercise. It calls in play the muscles of the chest. Infants should not be carried, or forced into an upright position ; as that might cause injury, or deformity to the soft, and imperfectly ossified bones. Much jumping, or dandling them is injurious ; and they should never be induced to walk *early* by teaching, or any artificial means, for the same reason. Nature is the child's best guide. It is better to keep children back, than to urge them on in this respect. If there is any debility of the system to retard them, *the remedy*, by all means, is Rush's Sarsaparilla and Iron.

Vomiting of Milk. Acidity. Flatulence.

If babies nurse too much, nature provides a remedy by throwing up the needless food, and no mischief results. But when this changes into vomiting of mucus, or watery fluid, or even bile, it needs a remedy. In the first place prevent their being overfed, especially with sugar; or spoon-fed, which is often done to prevent crying. If the above symptoms continue, and especially if there is flatulence, constipation, and bad smelling diapers, give small doses of Rush's Sarsaparilla and Iron; six to ten drops, every eight hours, would be suitable. The mother, if nursing, *should also take it.* Such troubles often come from the mother being Dyspeptic. See Dyspepsia. The nursing mother should remember that her babe is still a part of herself.

Crusted Tetter of Infants, Crusta Lactea.

This is a crusted and scabby eruption on the face, cheeks, and foreheads of babes. It begins with clusters of small whitish pimples, on a red ground. The crusts and scabs form in these. This disorder shows a Scrofulous tendency in the system; and no time should be lost in eradicating it. This may be done by the use of Rush's Sarsaparilla and Iron, which the mother should also take if she is nursing.

Bowel Complaints of Infants.

A healthy child at the breast soils on an average about five diapers in twenty-four hours. They may be considerably more frequent, however, if healthy-looking and smelling. But if they become green, and yellow, and watery; or brown, and white, and frothy, as if fermented; or if bad smelling, or attended with pain, something should be done. If the cause can be ascertained;—such as irritating physic, and herb teas, stuffing with improper food, fright, and exposure to cold;—these should be avoided; and if the mother is *Dyspeptic*, (see Dyspepsia,) she must remedy that. *Chamomile*, (page 151) is the most valuable remedy for diarrhœa, as above described, in restless, peevish, and weakly babes, or during teething. Put two drops in a tumbler of water, and give a teaspoonful, after each loose diaper. When diarrhœa occurs, at the time of weaning, from sudden change of food, and *with attacks of vomiting*, give Ipecac, in the same way as Chamomile; and also after each attack of vomiting. In all cases of loss of strength and flesh, weakness and pindling, paleness, or bloating, especially with signs of humor in the mother or child, the health of both will be rapidly restored by taking Rush's Sarsaparilla and Iron.

Eruptions and Humors of Infants.

These are of various kinds, occuring on different parts of the body, sometimes called Erysipelas; but also known by many other different names, such as Tetter, Ring-worm, Heat Spots, or Prickly Heat, &c., &c. All such humors show a tendency in the system to Scrofula; (which see,) which should be promptly rooted out by the use of Rush's Sarsaparilla and Iron. If the child is nursing, the mother should not fail to take the same remedy at the same time, and will find herself amply repaid by her returning bloom, and vigor, and flesh, and strength.

Teething. Dentition.

About the fifth or sixth month this begins. If due attention has been paid to the rules for bathing, exercise, diet, and medication laid down in this book, there will be but little suffering. During teething there is always a tendency of blood to the head, which may be guarded against by keeping the head cool; and particularly by not stifling the child's head when sleeping. During dentition, the child is more restless than usual, especially at night; has flashes of heat, alternating with paleness; gums hot, and swollen; some difficulty in suckling, bowels relaxed; and and there is more or less general irritability and fretfulness. The remedy is *Chamomile*, (151.) Put two drops in a glass of water, and give a half-teaspoonful every six hours.

Worms.

It is often supposed that children have worms, when the disease is in the blood. They are pale, weak, and sickly, and require Rush's Sarsaparilla and Iron to cure them, in nine tenths of the cases. The only sure sign of worms is to *see* them. For the long round kind, buy ten grains of Santonine, and grind it up in a teaspoonful of white sugar, and make ten equal powders; of which give one every other night, to a child four years old; give more or less according to age. This will kill, and bring away the worms, if any exist. Injections of lime water will kill thread worms; but do not use them when there are piles.

14

FEMALE COMPLAINTS.

The most lovely ladies may fall a prey to diseases peculiar to their sex. They are entitled to our warmest sympathies and most active exertions.

These disorders most frequently arise from humors in the blood, which settle in the parts affected; but also may be caused by displacements of internal organs.

All causes which weaken these parts, will draw to them any unhealthy humor in the blood; producing tumors, discharges, and ulcerations, and suspension of the healthy functions. Most cases will be cured or benefited by the remedies here recommended; but there are many cases which require our most attentive care to effect a cure. It is one of our most pleasing remembrances, to recall the hundreds of lovely women who have been, by our treatment and remedies, restored to charming health, with all its delightful enjoyments.

Ladies wishing to consult the editor, can do so with confidence, even in the worst cases, and free of charge. Address, at page 44.

Chlorosis. Green Sickness.

The suitable age for the commencement of menstruation, or the regular monthly sickness of healthy females, usually called being "unwell," is about fourteen; and it continues to about forty-five, in our climate. When this flow does not make its appearance in young females, we generally find a state resembling *anæmia*, or *consumption of the blood*, which see. We find a depraved appetite; pale, unnatural complexion; eating chalk, slate pencils, and the like, and general debility. There is constipation; cold hands and feet; flatulency; dyspeptic signs (see *Dyspepsia*); and sometimes spitting of blood, hurried respiration, and apparent signs of a decline. The remedy for this condition, and an unfailing one, is Rush's Sarsaparilla and Iron. Take it after each meal, and at bedtime; dose, two teaspoonfuls. If there is constipation, move the bowels twice a week with Rush's Pills. 210

SUPPRESSION OF MENSES. AMENORRHŒA.

This may occur at any time after the first appearance of menstruation, from wet feet, taking cold, and numerous dyspeptic and other causes, some of which are mentioned in the last article. *It is one of the first signs of pregnancy;* and generally a pretty sure one. *The remedy,* when suppression is not caused by conception, is Rush's Sarsaparilla and Iron, which will rarely fail to bring about the proper change by the next period.

In obstinate cases, take Rush's Monthly Remedy, according to the directions with the bottle; using the Sarsaparilla and Iron at the same time. For the proper use of Rush's Monthly Remedy, see that article.

PROFUSE MENSTRUATION. MENORRHAGIA.

When too much of the vital fluid is lost in this way, or where the flow returns too soon, or lasts more than four or five days, the system becomes rapidly weakened, and *anæmia or consumption of the blood* (which see), may result. The remedy for profuse menstruation is tincture of hamamelis (page 151). Put five drops in a tumbler of water, and take a table-spoonful every hour, until improvement begins; after that, every three hours. Take it every three hours for a day or two before the expected return. Where a state of *anæmia* has already become established, take the remedy recommended for that disorder, page

PAINFUL FLOW. MENSTRUAL COLIC. DYSMENORRHŒA.

This trouble usually occurs in unmarried ladies; and marriage is sometimes a cure. There are a variety of medicines, but their success depends upon an exact adaptation to the case, which cannot be expected in domestic practice. This disorder is also sometimes due to some physical obstruction, as at the os uteri, or mouth of the womb; and requires the very best treatment for its removal. For information how to obtain further advice, see page 44. Dysmenorrhœa is frequently attended by dyspepsia, or anæmia; read those articles.

WHITES. LEUCORRHŒA.

This is one of the most common of all female complaints, and embraces every variety of flow except red. If allowed to continue, it will bring about a permenent weakness of the female organs; giving rise to prolapsus, or falling of the womb, barrenness, polypus, and even cancer. The *best* remedy is Rush's Restorer. Put as much as can be held on a new cent into about a tumblerful of soft water, and use it very freely as an injection, three or four times a day. Prepare it freshly every time. Any kind of syringe can be used, but the best kind is the new rubber sort, which operates by gentle squeezing; no lady should be without one. Ladies using this remedy (the Restorer) will have the advantage of being able to use the same thing as a preventive, *if so desired.* (See *Prevention of Pregnancy.*)

Both for whites, and prolapsus uteri, or falling of the womb, we can not say too much in favor of sitz-baths. They should be taken at a temperature of about 65 degrees; one bath on going to bed, for twenty or twenty-five minutes, and, in severe cases, one bath at about 10 A. M., or 3 or 4 P. M., as may be most convenient. If only one daily bath can be taken, let it be at night. It will soon be found a great source of comfort and relief. The morning sponge-bath should not be neglected. See page 13 and 30. Baths can always be indefinitely continued, so long as found beneficial. *Ladies should remember that sitz and sponge-baths are to be continued just the same during menstruation, without fail.*

In all cases of Leucorrhœa, we may have general debility; paleness, and lack of blood; want of appetite, and other dyspeptic symptoms, as a result. (See *Dyspepsia* and *Anœmia.*) In all such cases, Rush's Sarsaparilla and Iron will prove a sovereign remedy, and should be used at the same time.

FALLING OF THE WOMB. PROLAPSUS UTERI.

This is very frequently the result of too early sitting, or walking, after confinement; also long-continued leucorrhœa, and all debilitating and relaxing causes. It may be known by a feeling of bearing-down pain, made worse by standing and walking, and by a sensation of dragging at the loins. Whites frequently accompany it, and may be cured at the same time. The remedy is Rush's Restorer, which is to be used as recommended for leucorrhœa, (Read last article.) In almost all cases of Prolapsus the system will be found laboring under general debility and poverty of blood, or dyspeptic troubles, the remedy for which is Rush's Sarsaparilla and Iron.

POLYPUS UTERI. BLEEDING TUMOR OF THE WOMB

This sort of tumor grows from the mouth of the womb, and may be known by its constantly giving rise to a colored flow   The sufferer says she is *unwell all the time*   She also has bearing-down; and in consequence of the loss of blood, becomes pale and bloodless, and suffers from dyspepsia and anæmia (which see). To restore her to health a surgical operation will be necessary to remove the polypus. But for the dyspepsia and poverty of blood, the remedy for those complaints can be used either before or after an operation. It is Rush's Sarsaparilla and Iron. Ladies wishing further information, see page 44.

HYSTERICS, HYSTERIA.

This disorder comes on with paroxysms of tears, crying aloud, difficulty of breathing, nausea, and palpitations, preceded, by anxiety and depression of spirits, and by pain in the left side, which seems to roll upward till it reaches the throat, when it feels as if a ball was lodged there, and a feeling of suffocation, stupor, and insensibility, and spasm of the jaws may follow. The body and limbs are agitated; there are fits of laughing, crying, and screaming, and foaming at the mouth; relief ensues with eructations, and frequent sighing and sobbing; followed by a soreness over the whole body. Hiccough, or hiccups, is frequently very troublesome. In many cases there is violent pain in the back, or at the pit of the stomach. Hysteria may resemble very many other diseases, and have many or all of their symptoms. The causes are scrofula, poverty of the blood, bad diet, sedentary habits, and exciting and depressing passions. It often arises also from chlorosis, amenorrhœa, or menorrhagia. (See those articles).

When an attack of hysteria comes on put five drops of pulsatilla (see page 152) in a tumbler of water, and give a table-spoonful every half-hour. If pulsatilla is not at hand, use half a teaspoonful of camphor in the same way. This disease can be radically cured by removing the causes, and by taking Rush's Sarsaparilla and Iron, until the system is renovated, and the general health is restored. If there are whites, or prolapsus, use Rush's Restorer at the same time. (See *Whites*.)

BARRENNESS.

This is generally caused by scrofulous humors, want of amativeness, anæmia, or consumption of the blood, (see *Anæmia*), defective nutrition, habitual costiveness, and dyspeptic troubles. (See *Dyspepsia*.) We now have a very sure remedy in Rush's Sarsaparilla and Iron. Take two teaspoonfuls after each meal, and at night. This should be continued patiently even for six months or a year, until the desired result is obtained. Marital indulgences should only take place during the first week after menstruation, as that is the most fruitful way. If there are *whites*, or *falling of the womb*, conception cannot be expected until they are cured. (See *Whites*, &c.) After connection, the lady should lie quite still for two or three hours, *at least*, that fecundation may take place. No barren lady, or one who lacks enjoyment, will ever use Rush's Sarsaparilla and Iron with regret. It is veritably the elixir of Cupid and Venus.

Pregnancy. General Observations.

The period of child-bearing is one of peculiar interest, and usually regarded as subject to many discomforts. There is, however, no good reason, if the laws of health are observed, why the whole period should not be passed in comfort and good health. The expectant mother should, therefore, remember that the propriety of a regular and systematic course of life devolves upon her with double force. Since every neglect of the laws of health, upon her part, is frequently visited with fearful energy upon her yet unborn infant. (See *Laws of Health.*) Every pregnant female should take some exercise every pleasant day in the open air; and the best is walking. Pure air is one great point to be gained, and that is why exercise out of doors is to be sought. Too much exercise in doors is to be avoided, as inducing want of sleep, over-fatigue, and waste of the vital powers, now so necessary to a new and double life.

The clothing should be loose, and particularly *all pressure* by tight-lacing, or otherwise, on the abdomen and breasts, should be avoided. From pressure on the breasts, retracted and sore nipples, obstruction of the milk, abscess of the breast, and even tumors and cancers may result. Pressure on the abdomen may cause loss of natural elasticity, by which permanent abdominal deformity may result, from non-contraction after confinement. Misplacement, club-feet, and often malformations may also result to the fœtus, or unborn child, productive of long and dangerous labors. *Tight garters* are injurious from the pressure on the blood vessels, causing swelled veins, which sometimes become very painful and troublesome. Garters should be placed just *above* the knee, as is the custom in Europe. The symmetry of the calf of the leg is frequently spoiled by the pressure.

The rules for diet, as laid down in this book, should be strictly observed; the amount of food should be moderate; and particularly coffee, pork, and strong tea should be shunned

The mind has its effect also upon the fœtus; and the

cultivation of mental improvement and harmony, as well as health of body, will powerfully conduce to the future well-being of the unborn babe, which the expect·ant mother carries so near her heart. All mental excitement, *especially religious*, as well as late hours and fashionable dissipation, should be carefully avoided Husband and wife should be very temperate, and almost abstient

Unsightly objects should be carefully avoided, as possible to affect, or even mark, the child ; and agreeable subjects cultivated ; such as music, painting, views, and engravings ; and scientific, manufacturing, agricultural, domestic, and harmonial improvement.

Despondency of mind, low spirits, and uneasiness about the future, are very common during pregnancy. When this is accompanied by any dyspeptic signs (see *Dyspepsia*), as it usually will be, *the remedy* is Rush's Sarsaparilla and Iron. Take a teaspoonful after each meal

Menstruation during Pregnancy.

This is very rare, especially after the second month, though there is a popular opinion to the contrary. If it should in any case continue during the latter months, the remedy is hamamelis (see page 151). Take one drop every night on going to bed. If dyspeptic signs are present (see *Dyspepsia*) take the remedy for that.

Morning Sickness.

This, with nausea, vomiting, and heart-burn, are the most common and frequently most distressing symptoms of pregnancy. They are most troublesome on first rising. They generally disappear at about the period of quickening ; that is, between the fourth and fifth months, but not always.

The treatment of this derangement by Rush's Sarsaparilla and Iron is at once simple, prompt, and efficacious. Take a teaspoonful after each meal, on going to bed, and on awaking in the morning. (Observe *Rules for Diet*, page 21.)

COSTIVENESS IN PREGNANCY.

Rush's Pills are better than any other cathartic pills; but the proper remedy to effect a cure, and at the same time strengthen the system, is Rush's Sarsaparilla and Iron. Take a teaspoonful after each meal, and at bedtime. This will be almost indispensable if coffee is used.

TOOTHACHE IN PREGNANCY.

This is frequently very troublesome, and may occur in quite sound teeth. Such should not be extracted, but decayed ones should be. An aching tooth may frequently be relieved by holding very cold water in the mouth, frequently changed. It usually depends upon scrofulous or humory taint; and when that is the case, it shows the necessity of purifying the blood by taking Rush's Sarsaparilla and Iron; when the toothache will not fail to disappear.

SWELLED VEINS. VARICOSE VEINS.

This is caused by the pressure of the gravid womb. The remedy, is abundant bathing with cold water, and bandaging from the feet upwards with a gentle pressure, or a laced stocking may be worn. When this complaint is severe, the limbs must be supported on cushions, or pillows, or the patient may lie on a couch or bed.

PAINS IN THE BACK IN PREGNANCY.

These are sometimes very severe, are situated in the lumbar region, and may prevent sleep at night. A very good remedy will be found in the sitz-bath, taken for twenty or twenty-five minutes before going to bed. The water should be at a temperature of about 65 degrees. (See *Water Treatment.*) This remedy is very valuable in pregnancy, strengthening the whole uterine region.

MISCARRIAGE. ABORTION.

Before the seventh month of pregnancy this term is used; after that it is called *premature labor.* When it has once occurred, it is much more likely to take place again. The third or fourth month is the most common period, and it is more liable to be attended with profuse and dangerous hæmorrhage, or loss of blood. The previous signs vary very much; sometimes the discharges and pain are very considerable; at others very slight. The causes are sudden mental emotions, great physical exertions, mechanical injuries, a luxurious mode of life, powerful cathartics. neglect of air and exercise. Abortion is preceded and attended by the following signs: A slight chill, followed by fever, with bearing down; severe pain in the abdomen, drawing and cutting pains in the loins, like those of labor; discharge of mucus and blood, sometimes bright red, at others dark or mixed with clots. The fœtus is generally lost during this discharge, which may continue for hours, placing the sufferer in peril from loss of blood. When the pains increase in intensity, and the regular throes become established, with efforts to dilate the mouth of the womb, abortion is highly probable.

As a prevention of this affection, the sitz-bath on going to bed, for twenty minutes, at 65 degrees, is to be highly recommended. (See *Water Treatment*.) All the causes mentioned should be avoided, and if there be any signs of scrofula, dyspepsia, whites, or falling of the womb, the proper treatment laid down under those affections should be made use of. In case some of the signs come on, with slight pains and flow, put five drops of hamamelis (page 151) in a tumbler of ice-water, and give a table-spoonful every half hour. At the same time observe perfect rest in bed, and apply linen cloths kept wet with ice-water, to the lower part of the bowels, and covered with a dry towel. Should this prove of no avail, or only partially successful, after three hours, give first pulsatilla, next chamomilla, in the same way. Those who prefer stronger doses, may procure a half an ounce of elixir of vitriol, and give twenty drops in sweetened ice-water every hour. The immediate danger being over, perfect quietude should be observed for some days.

When any accident has occurred, as a fall or blow, or an operation has been recklessly attempted, give arnica in the same way as hamamelis, and use the same to wet the cloths applied, as before directed.

We cannot here too strongly condemn the practice of procuring abortion by surgical operations, or poisonous drugs. Such a practice is not only likely to prove fatal to future hopes of offspring, if ever desired, but to cause polypus, tumors, or even cancer of the womb. The whole matter of procuring abortions has now become wholly needless with careful persons, since there is in Rush's Restorer and Preventive an infallible means of preventing conception, while health is promoted at the same time. (See article on *Prevention of Pregnancy*.)

THE BREAST.

Young mothers frequently find difficulty in nursing. This may arise from pressure exercised upon the breast and nipple in childhood, or by the pressure of corsets in after life. Any pressure of this kind should be avoided, as it may so far destroy the structure of the milk tubes as to render the process of milk production very painful, or wholly impossible. Sore nipples, which are often a source of great suffering, may be prevented by bathing them, once or twice a day, with nice whiskey, or Medford rum, for two weeks before expected confinement. The effect of this is to harden the skin, and prevent cracks. If the nipple is short or retracted, so that the babe cannot easily hold it, it may be caused to protrude properly by a nipple shield of wood, or glass, worn before confinement. If this trouble should continue, a rubber and glass pump, by exhausting the air over the nipple, will cause it to elongate, and give the babe sufficient hold. The same instrument is very convenient for drawing the breast, in case the babe does not need all the milk, or from any cause is unable to nurse.

It is very important, where there is a plentiful supply of milk, that the breast should be nursed or drawn every three or four hours, *and never suffered to become hard*, as that will cause obstruction and inflammation of the milk glands and tubes, and *an abscess* or BRO-

KEN BREAST. Careless nurses, *by allowing the breast to remain undrawn all night,* in nine cases out of ten, *cause* this distressing misfortune. It is a fault of *omission,* and a great one, to allow the breast to remain undrawn all night, especially during the first few days of milk production. In case, however, this accident *has* happened, and a hard and tender lump in the breast, when taken between the thumb and finger, is found, there is great danger of abscess. In such cases put ten drops of arnica (see page 157) in a basinful of cold water, and apply a napkin, kept well wet with it, to the hard and tender breast Change this often, and a dry flannel may be applied outside ; at the same time draw the breast every two hours, as much as can be done without much pain ; and give Rush's Lung Balm, and Belladonna (see page 151), as follows : — ten drops of balm ; then, in one hour, put five drops of Belladonna in a tumbler of water, and give a table-spoonful ; and so on once an hour, alternately, until improvement takes place. If no Belladonna is at hand, use the balm and wet compresses alone. As soon as the hardness and tenderness begins to disappear the compresses may be left off at night, and the balm given once in three hours.

This plan, if followed up, will save many a *broken breast :* and will always diminish the size of the abscess, *prevent its attacking the other breast,* and bring it to a head quickly. After pus or matter has formed, which may be known by *a chill,* it should be *soon* opened ; as this will save much loss of the substance of the breast. The folded wet napkin, or compress, is much better than a poultice, and has the advantage of cleanliness. An abscess of the breast, as usually treated, is attended by long and profuse discharge of pus, or matter, and results in much debility and loss of appetite and strength ; to prevent this, or remedy it, use Rush's Sarsaparilla and Iron. (See page 47.)

Deficient milk sometimes is a serious trouble. The quantity may usually be much increased by *eating raw apples,* which, are besides, a most healthy and nutritious fruit. They may also be eaten cooked. A deficient secretion of milk, however, usually shows anæmia or poverty of blood, and depends upon a lack of tone in the digestive organs. In that case the remedy for anæmia (page 181) and dyspepsia (page 164) must be used.

Suppressed secretion of milk, or stoppage of the flow of milk, may result from a variety of causes, of which more or less *fever* is the most common ; when that is the case put five drops of aconite and the same of pulsatilla (page 151) in separate tumblers of water, and give a table-spoonful of aconite, and in one hour the same of pulsatilla, and so continue until the fever subsides ; then give only the pulsatilla until the secretion of milk is restored.

MILK FEVER.

The natural flow of milk does not require medical aid for its regulation. Still many females suffer slight uneasiness during the first flow, and when the following signs are present the affection is called *milk fever :* Thirst, shivering, and heat, terminating in perspiration, pulse variable, pain in the back, extending to the breast, disagreeable taste in the mouth, oppressed breathing, anxiety, headache, and diminution of the milk and lochial flow. The feverishness is worse in the evening, and passes off by perspiration in the morning. This succession of signs may recur on the following day, but nature is usually able to restore the equilibrium of the system. If, however, this does not take place, or if the fever runs high, put five drops of aconite in a tumbler of water, and give a teaspoonful every hour, until perspiration ensues. Then give Rush's Lung Balm, ten drops every three hours, until the milk and lochial flow are restored. It should be remembered that neglecting to put the child to the breast sufficiently early is a frequent cause of severe milk fever. The milk which has been secreted, not being drawn out, is reabsorbed into the blood, and causes the fever.

False Labor Pains.

These are brought about by various causes, and sometimes precede labor but a few hours, but in many cases come on some days, or even weeks, before delivery. They differ from true labor pains in being irregular in returning, and disconnected with contraction of the womb. They are chiefly confined to the abdomen, or belly, with sensitiveness to touch and movement, and the pains do not increase in violence as they return. The period of pregnancy is one important guide to distinguish the true from the false. It should be remembered that the natural full period of pregnancy is forty weeks, or two hundred and eighty days, which exceeds nine calender months by nearly a week. This time is to be reckoned invariably from the *end* of the last monthly flow. If this time is not nearly expired, false pains are more probable. A physician distinguishes false pains from true by examining the mouth of the womb, which remains quite closed in case of false pains. False pains should be quieted by small doses of Rush's Lung Balm, of which give five drops every two hours. If there are symptoms of dyspepsia present (see page 164), or of anæmia (see page 181), do not fail to take the remedy there recommended.

Natural Labor.

This should occur at the end of forty weeks from conception, and the time should be reckoned from the end of the last monthly period. If the directions for diet and laws of health, given in this book, have been followed, the contractions of the womb will be regular and effective, and the whole process will rarely continue more than twenty-four hours, and usually not more than twelve, or even less, in all well-formed females. We may observe that among savages, where plenty of fresh air and exercise, bathing and unstimulating diet, and loose clothing are the rule, labors are comparatively free from pain and danger.

The beginning of natural labor may be supposed when regular pains have set in, *which increase progressively* in severity; and this will be still more probable, if a slight flow of blood, called *a show*, has occurred. After this time, even if it has been delayed so long, the physician should be sent for. He is the only proper person (well educated female physicians excepted) to take charge of such cases, but as a physician is not always pro urable, every married lady should know the ordinary duties to be attended to, and we shall therefore briefly state them. The head of the child is usually first born. The next pain will probably expel the body and limbs; if compelled to assist without a physician, as soon as the head is born, see that the babe's head is raised a little, so as to clear the mouth and nose from the bed, and give the little innocent a chance to cry. Wait a few minutes, say five, after the whole body is born, and take a piece of strong tape, or twine, and tie a very firm hard knot round the navel string of the child, called the *funis*, which will be found at the babe's navel, and forms the connection with the mother. This knot should be about two inches from the babe's navel, and should be drawn very firmly, and tied very well, with at least three hard knots. As soon as this is done cut off the *funis* about an inch outside your knot. As soon as cut examine the cut end, next your knot, and if there is any bleeding apply another cord and knot next to the first one. The *funis*, or naval-string, is generally about twenty inches long, about the size of the little finger, and is of a gristly substance, containing three large blood-vessels. The other end grows from the *placenta*, or *after-birth*. Hence after the babe has been separated, if the after-birth has not been expelled by a return of the pains, after waiting a half hour it is common to pull very gently by the *funis*, in order to bring down the after-birth, and complete the delivery. Put the other hand, at the same time, on the lower part of the bowels, over the womb, and make a gentle kneading pressure, also telling the lady to bear down. The pulling by the *funis* should not exceed what can be done by *one hand*, without winding the navel-string about the fingers. A lady, caught without a physician, may safely wait twenty minutes before separating the child, which should be turned on its *right* side, as it lies near the mother, with its mouth and nose clear. An hour or two may be allowed to elapse before taking away the after-birth. A physician is entitled to his fee, if he arrives before the after-birth is born.

The babe having been separated as above described, in taking it up be careful not to let it fall, as it is very slippery. (For care of the infant see page 206.) Any lady or husband called upon to assist under such circumstances should remain quite calm, and if this book is at hand read over these directions, by following which you will do your best. For the first two or three days, especially *for the first few hours*, the mother should be very careful not to sit up, or particularly to *stand up*, even for a moment, *as that might cause a dangerous flow of blood.* She should be moved, and *made comfortable, while constantly lying down.* A cup of black tea or some oatmeal or arrow-root porridge, or gruel, may be taken as soon as the mother feels any need of refreshment, or within an hour or two after delivery. Among the cooking receipts of this book may be found some very suitable for the first few days. We would here repeat that every mother who tries to sit up or go about in the first eight days, *in order to be smart,* runs the risk of *falling of the womb as a result.*

The binder is a bandage of cotton or linen, eight or ten inches wide, and long enough to go once and a half round the body. It should be applied round the uterine region, with moderate firmness, soon after the birth of the child. Its object is to support the relaxed organs, and it should only be worn a few days in most cases. A pillow-case answers for a binder, and it is better for the physician to apply it.

Flowing or Flooding after Labor.

This is the result of the womb not properly contracting after delivery when blood flows from the open mouths of the uterine vessels, where they join the after-birth during pregnancy. The blood flows in gushes, externally, or comes away in clots, or the womb may even fill with blood, and *faintness* come on from the bleeding. If any unusual flow ensues, *with faintness*, in the absence of your physician, remember that the only safety consists in a permanent contraction of the womb. Place the hand upon it. It should be felt as a hard tumor about half the size of the babe's head. To compel it to contract begin a gentle kneading of it with the open hand; if that brings on pains so much the better, *there is always safety in after-pains.* In the mean time apply a folded towel directly over the womb, previously wet with ice-water, or very cold water having in it a teaspoonful of Rush's Restorer (see page 231), or ten drops of arnica (see Appendix page 151). Prepare a basinful of it, or use water alone. Cover the cold wet towel with dry flannel, and keep the towel wet and cold. At the same time keep the feet well warm, and even hot, with hot bricks or by putting them in hot water. This is important. The cold compress will rarely fail to stop the bleeding; but you must *get, and keep* the feet warm, to prevent taking cold. When the bleeding is stopped the compress or towel may be removed; do not let it remain and become hot and dry. You may give, as a medicine for the flow, five drops of Hamamelis in a glass of water, a table-spoonful every half hour. If there is *faintness take away the pillows* from the head, and give a little warm currant-wine or spirits and water, by teaspoonfuls, until the faintness passes off.

Some nourishment may be given of toasted bread covered with new milk, or cream and water.

We would here add that all the bathing about the mother, in making her comfortable, and afterwards, should be done with tepid water having in it a teaspoonful of Rush's Restorer, which is best; if not at hand, ten drops of arnica (see page 151). *This will remove all soreness, and prevent puerperal fever.*

AFTER PAINS.

These are the efforts of the womb to clear itself of the blood, or clots, which have collected in it after delivery. They are a *sign of safety* from *flooding* (see last article), and close up the bleeding inside of the womb. After first labors they are generally very light, but should they at any time become severe, they may be relieved by laudanum, of which give thirty drops, and if not quieted in three hours, twenty more may be given; but it is better to do without.

Treatment after Delivery.

A gentle sleep is usually the first and best restorative. To favor this, the room should be kept dark, and perfectly quiet, and not too warm, not over 65 or 70 degrees. (See *Nursing*, page 34.) If sleep is prevented by tossing and restlessness, give a teaspoonful of Rush's Sarsaparilla and Iron, and if this does not succeed, put five drops of coffee in a glass of water, and give a teaspoonful every hour. If there should be a *chill* at any time, or if feverish signs arise, give aconite in the same way (page 151). It is customary to give something to move the bowels after two or three days. We doubt the propriety of this. If anything is used give *one* of Rush's Pills every night until a motion takes place; but it is better to wait four or five days even, and then give a large injection, or more than one, of tepid water. This will rarely fail to have the desired effect, without medicine.

Perspiration after Delivery.

The increased perspiration, which takes place after child-birth, serves as a substitute for motions of the bowels; consequently its sudden suppression is followed by a bad result, and sometimes by puerperal, or child-bed fever. To remedy this, give a half a teaspoonful of Rush's Sarsaparilla and Iron every hour, and keep the feet warm; take each dose in a large tumblerful of water, the whole to be taken as rapidly as convenient. This will powerfully incite perspiration. As soon as it begins, leave off the medicine.

Debility after Delivery.

This may arise from excessive perspiration, or from want of appetite, child-bed fever, loss of blood, or too great lochial flow, and other weakening causes. It will be very promptly removed by taking Rush's Sarsaparilla and Iron (see page 47). It should be continued until health and strength are entirely restored.

Lochial Flow. The Cleansings.

The flow after delivery varies very considerably. Sitting up too early, rich and stimulating diet, keeping the room too warm, mental emotions, and irritating causes generally, may cause it to be profuse and long continued. If the flow is bloody and profuse after nine days, put five drops of hamamelis (see page 151) in a tumbler of water, and give a tablespoonful every two hours. At the same time take Rush's Sarsaparilla and Iron (see page 47), and continue it until the health is entirely restored. Omit the hamamelis as soon as improvement takes place. If the flow is profuse but not bloody, use Rush's Restorer (see page 231) by putting a teaspoonful in a pint of water, and using it very freely as an injection three times a day. A profuse flow may degenerate into *whites*, when the treatment of that disorder will be required (see page 212). When the lochial flow is *suddenly suppressed*, from cold, or other causes, fever may set in. To avert this, put five drops of aconite in a glass of water, and give a table-spoonful. In one hour give pulsatilla (page 151) in the same way, and so continue alternately, with one hour between the doses. As soon as slight perspiration begins, or the flow reappears, give only the pulsatilla until it is all right again.

If the discharge should at any time become sanious, fetid, or offensive, use Rush's Restorer as just recommended.

Child-bed Fever. Puerperal Fever.

This is most to be dreaded of any of the common accidents attending confinement. But if the directions given in this book in regard to the *Laws of Health and its Preservation*, and the precautions in regard to *diet, nursing, natural labors, and treatment after delivery*, we venture to say that there would be very few cases of puerperal fever. "An ounce of prevention is worth a pound of cure." Every lady expecting confinement, therefore, should read all these articles, and that on *the breast*, and govern herself accordingly.

This fever consists in an inflammation of the peritonæum, or glistening membrane, which lines the cavity of the abdomen or belly, and also covers the womb, vagina, bladder, intestines, and all the abdominal organs. It is dangerous, from its tendency to spread through the whole extent of this membrane. This fever is caused by taking cold, suppression of perspiration or the lochial flow, by injuries done to the womb, or vagina, *by attempts to procure abortion*, or by difficult labors, by too rich and stimulating diet during confinement, by violent purgatives, and by other irritating causes.

The signs of puerperal fever are : A chill, generally severe, *with cold feet*, then flashes of heat, and the signs of inflammatory fever (see page 155). Perspiration and the lochial flow are suppressed. We find *great tenderness of the abdomen, the mother lying with her knees drawn up, and unable to bear the slightest pressure*, even of the bedclothes. The face looks pinched and anxious, and the pulse is very rapid and inflammatory. There is severe pain in the lower bowels, *made worse by pressure*.

This is not a disease which can be treated to advantage in domestic practice, and we have only introduced it here for the sake of pointing out the necessity of avoiding it, and its causes. And we would mention, moreover, that many of the best physicians now believe that it often has resulted from contagion at the hands of the physician himself. This contagion consists of particles of morbid matter, from *Erysipelas, other cases of puerperal fever, and dead bodies dissected*, and is conveyed directly by the hands of the physician. In one of the large Paris hospitals, while the editor was attending there, the mortality was reduced more than fifty per cent. in this fever by the simple expedient of causing all the medical men to dip their hands in a disinfecting solution, and use a nail brush with soap, before passing from diseased persons to attend cases of labor. We advise our lady readers not to employ a physician in confinement, who is known to be attending Erysipelas patients, or who has now on hand, or lately has lost cases of puerperal fever. It has already been mentioned that *attempts to procure abortion by means of pointed instruments* often cause this disease. The inflammation, beginning at the womb, spreads rapidly and causes death. There is no excuse for married ladies resorting to this dangerous method, since they have in Rush's Restorer and Preventive a sure and healthful method of avoiding such predicaments (see page 231). We have mentioned that a *chill*, with cold feet and hands, is one of the first signs of child-bed fever. A recently delivered woman who has a *chill* never knows what accident may or may not be her lot. She should immediately have her feet and hands got warm, and kept warm, by putting them in hot water, and then applying hot bricks, rolled up in a damp cloth. At the same time put five drops of *aconite* (see page 151) in a glass of water, and give a table-spoonful every hour, and drink freely of cold water, say nearly a pint an hour; all which will be strongly promotive of perspiration. As soon as this sets in, and the lochial flow returns, the attack will usually be thrown off, and the remedies may be discontinued. The above treatment can never do harm ; but should all the signs previously mentioned set in, no time should be lost in procuring the best medical attendance. Until a physician arrives, put five drops of belladonna (see page 151) in a glass of water, and give a table-spoonful one hour, and the aconite before directed the next, and so on. Examine the feet frequently, and keep them hot.

The above would be appropriate treatment for at least one day, and better calculated to throw off an attack, than giving bulky drugs.

Sore Nipples.

This has been already mentioned in speaking of the *breast* (see page 219); it was there stated that a *nipple shield*, of glass or wood, should be worn, in case of retracted or compressed nipples, before confinement, to cause them to protrude, or grow longer, by taking off the pressure of the dress, there being an aperture in the shield to receive the nipple. In case of sore nipples, this should be worn during the intervals of nursing, taking care to keep both the nipple and surrounding breast *wiped quite dry*, to prevent rawness and excoriation; and to keep the shield from adhering to the skin, as it might do if moisture were permitted to collect under it.

If the nipples have contracted cracks or fissures, it will be best to draw the breast for a day or two with the rubber and glass breast-pump, while the cracks may have a chance to heal under the application of *arnica court plaster*. (See page 151.) The infant may be, at the same time, nursed from a bottle, which should be kept perfectly sweet. In case it seems impossible to restore a nipple to a sufficient length to enable sucking, an artificial nipple may be worn, made of a heifer's teat. These are sold by most druggists. They may be employed at any time in case of sore nipples, while cracks are healing.

Mental Emotions Injure the Breast.

It is a well known fact that fright, or a *fit of anger*, may so poison the milk as to seriously injure, or even kill a nursing child. After such strong emotions do not give the breast to the child until a portion of milk has been drawn off, and all injury will be avoided.

Weaning and Declining Nursing.

Mothers declining nursing soon after confinement, and wishing to dry up the milk, should live as low as possible, and avoid juicy fruit and vegetables; at the same time put five drops of pulsatilla (see page 151) in a tumbler of water, and give a table-spoonful every three hours. This alone will often be sufficient to stop the secretion.

RUSH'S MONTHLY REMEDY.

The restoration of the suppressed function of menstruation has never been attained so safely, pleasantly, and effectually, as by the use of this remedy. It is in the form of an aromatic balsam ; and, unlike other nauseous compounds, in the market, is not unpleasant to take. It effects a restoration of the monthly flow, by bringing about a harmonious action of all the female organs ; without nauseating, purging, or any other disagreeable result. It is expressly adapted to obstinate cases, where other medicines have failed, and perfectly safe to be taken for this purpose under all circumstances ; at the same time with gentle action, being the most powerful remedy known , especially at the end of the first, or second suppressed period. It is greatly superior to all female pills. Ladies who suppose themselvs pregnant, should by no means take this medicine, as miscarriage would be the inevitable result.

Directions for using. Take thirty-five drops, in ordinary cases, every three hours, shaking the bottle, and gently warming it, in cold weather, before using it. In more difficult cases a teaspoonful may be taken for a dose ; but that quantity should not be exceeded. It is sold by most druggists, price 75 cents ; but may always be procured by sending 85 cents by letter to the proprietor of Rush's Remedies, Lowell, Mass., when the medicine will be forwarded free of charge. Observe that the letters A. H. F. are blown in the glass of each bottle ; all others are counterfeit.

RUSH'S RESTORER AND PREVENTIVE.

The prompt and effectual cure of female complaints, of ordinary kinds, such as Leucorrhœa, or Whites ; Prolapsus, or Falling of the Womb ; and other common weaknesses, is very readily effected by the use of this remedy. It acts by giving a new, and healthful tone to the female organs ; and by removing relaxed debility, and unhealthy secretions, and restoring the normal vigor, and elasticity. Rush's Sarsaparilla and Iron, can generally be taken at the same time with advantage.

It is also the most reliable remedy known for the prevention of conception, or pregnancy ; which it effects by destroying the vitality of the semen ; and at the same time has a most healing power upon the female organism.

Directions. Put as much of the powder as can be held on a new cent into a half pint of water, and inject it freely three or four times with a female Syringe. As a preventive, use it *immediately after* enjoyment. Prepare it freshly every time, and use it twice or three times a day, to cure female complaints. Any kind of Syringe will do ; but the best is the Rubber kind, which works with gentle squeezing with the hand. All druggists have them ; and usually keep this medicine ; price 75 cents a package ;

or it may be procured by sending the money by letter to the Proprietor of Rush's Remedies, Lowell. Mass., when the medicine will be sent free of charge. Observe the written signature of the proprietor, on every box.

For the cure of Gonorrhœa by this medicine, see that complaint, in Rush's Family Physician. It is the most effectual, and rapid remedy known, for that purpose, curing the discharge in a very few days.

THE LAWS OF MATERNITY.

The Healthful Prevention of Pregnancy.

This has always been regarded as a most delicate subject; and yet it is so much desired in thousands of cases, and calculated to do so much good, that we cannot forbear, in this book, intended for the benefit of the people, from making known, and defending the means of effecting this object. The whole course of human knowledge, improvement, discovery, and invention, since the beginning of the world, goes to show that whatever tends to the actual good of a portion of the human race, without injuring the remainder, may be fairly acted upon, under the general laws of the great Law Giver

Good children, are among the highest, and best of blessings. Nevertheless, there are thousands of cases in which, if it is not an actual sin against heaven, it is at least a positive crime against society, to beget offspring. Thus, when struggling parents bring forth from one to twenty children, who are almost sure to be victims of poverty and ignorance, they may esteem themselves fortunate if their children escape crime, and punishment.

Accordingly, it is evident, that with persons in humble or even moderate circumstances, a means for the prevention of pregnancy, admitting of full enjoyment, is a consummation most devoutly to be wished; for it is all nonsense to suppose, with some of our modern writers that any code of laws, or morals, can be formed that will keep the opposite sexes of the struggling classes asunder. In this respect nature smiles at human intervention, and insists on her prerogatives.

The writings of Political Economists, which would leave the rich alone, but are perfectly despotic as regards the poor, will have as little effect, as their arguments are incompatible with nature; for men and women will either get married, or do worse; in spite of all the theories of all the total abstinence writers in the world.

But then the evils of which those political economists complain, namely, the overgrowth of the poor in densely-populated countries, may be safely averted; and through a medium, too, that would be highly popular with the people if it were only known. No poor man or woman wishes to be overstocked with children; though of course, when they do come, they make the most of them; and therefore they would rejoice in an innocent means for preventing that effect, without interfering with their connubial enjoyments.

Nor are the poor alone concerned; but at least one third of the upper classes of all communities. Vast numbers of persons afflicted with hereditary diseases, such as scrofula, king's evil, &c., forfeit the pleasure of domestic life, because they conscientiously dread their misfortunes upon their offspring. And many females are of such physical organizations, that they cannot bring forth children without imminent danger of their lives, and are thus debarred from marriage, unless at the risk of self sacrifice. And again, ladies who are prone to conception (be their constitution ever so good, affection for their progeny ever so ardent, and their means ever so affluent), must regret to be kept in a continual state of pregnancy. I need scarcely add, that all the above, and many others for many causes, would rejoice in the means of avoiding the vexations and dangers alluded to. In support of these views read the following from the celebrated Robert Owen :—

" See what a mass of evil arises from illegitimate children, from child-murder, from deserted children, from diseased children, and even where the parents are most industrious, and most virtuous, from a half-starved, naked, and badly housed family—from families crowded into one room, for whose health a house and garden are essential. All these things are a tax upon love, a perpetual tax upon human pleasure, and upon health; a tax that turns beauty into shrivelled ugliness. Then comes the consideration— what a dreadful thing it is that health and beauty cannot be encouraged and extended—that love cannot he enjoyed without the danger of conception; when that conception is not desired; when it is a positive injury to the parties and to society. This circumstance has been a great drawback to health, strength and love.

" What is to be done to remedy this evil? There is something to be done; a means has been discovered, a simple means, criminal in the neglect, not in the use. The destruction of conceptions has been sought by acts of violence, by doses of poison, that injure, and sometimes destroy the mother to reach the foetus in her womb. This is dreadful, truly dreadful. Every village has its almost yearly cases of the kind. Hundred of infants are yearly destroyed at birth; some cases are discovered, but many pass undiscovered. We condemn and shudder at the infanticide of China and other countries; yet it is a question if infanticide ever prevailed in any country to a greater extent than in our own. Here, then, as in every other case of disease or other evil, *it is better to prevent than to* DO WORSE. Prevention pleases the mind of a woman at first thought: and once practiced, all prejudice flies, and approval must be the consequence. To weak and sickly females—to those to whom parturition is dangerous, and who never produce living or healthy children, prevention is a very great blessing. And it is also a real blessing in all other cases, where children are not desired. It will become the very bulwark of love and wisdom, of beauty, health, happiness, and virtue. If the question of love were thus made a matter of sedate and philosophical conversation; the pleasure arising from it, would be

greatly heightened, desire would never be tyrannically suppressed,
and much misery and ill health would be avoided. Parents
would explain its meaning, and its uses, and its abuses to their
children, at a proper age ; and all hypocrisy, and what is worse,
all ignorance on the subject, which lead to so many disasters,
would cease. We should soon see a much finer race of human
beings ; a much more chaste, and virtuous race than we now see.
Restraints on love operate precisly as they operate in cases of
excessive taxation; they destroy the revenue sought, and produce
the evils of smuggled, and mere disastrous intercourse.''

The very best, and *only safe, certain, and healthful means* yet
devised for this purpose, consists in the use of Rush's Restorer
and Preventive.

It should be used as directed in page 231. Directions also accompany each package of the medicine. Ladies using it for this
purpose will enjoy the advantage of strengthening the parts; and
removing soreness, relaxation, whites, and other weakness at the
same time. An instrument of the most approved construction
for the use of this remedy, with a supply of it sufficient to last
one year, will be forwarded, paid, by express, on receipt of $5.00 ;
address, as at page 44.

Mercurial Diseases.

We have already mentioned, (pages 94, 95,) that Mercury may
cause *Consumption*, and we there related the destruction to
animal life which followed the spilling of a large amount of
quicksilver in a ship at sea.

Calomel is the chloride of mercury, or quicksilver, and the most
extensively used, and therefore most injurious, of all mercurial
preparations, of which there are over twenty in all. It is in the
form of a tasteless white powder, and is often given to patients
without their knowledge.

The teeth are almost certain to be destroyed by this drug ; and
the great number of young persons, whom we see almost daily,
with teeth badly decayed, can confirm this.

The following are some of the disorders which Pereira, in his
great work on Materia Medica, the largest in the English language, ascribes to it.—Mercurial Fever ; Excessive salivation ;
Violent purging; Chronic Skin diseases ; Inflammations of the eye,
throat, and bones ; Enlargement of the glands; Ulceration of the
mouth ; Cachexia, or general wasting ; and Mercurial Palsy, and
other neuralgic affections We have seen in our private practice,
persons emaciated to a skeleton, *with both plates of the skull almost completely* perforated, the nose half gone, breaths more
pestiferous than the Upas, and limbs wracked with the pains of
the inquisition.

Modern Physiology and Pathology abundantly prove that no
chemical element should be introduced into the system, which
does not form a part of its tissues. We may add that all the
remedies recommended in this book, or used by the editor, con-
form to this law. Persons suffering with mercurial diseases,
and wishing information, or advice, may consult page 44.

SYPHILIS.

This disorder arises from impure intercourse, and makes its first appearance in the male, by a small pimple on the penis, which rapidly increases in size, and soon becomes an open sore. In most cases the syphilitic poison, or virus, is at the same time absorbed into the blood; and the whole system is infected simultaneously with the appearance of this venereal sore, or *chancre*. There may be more than one of these on the genitals; which, if left alone, or treated by mercury, may run together, and cause *fearful destruction* of the affected parts.

Not much pain is felt; and in the female, the infection may pass unnoticed for some days, or even weeks, until *very considerable swelling*, or large sores, attract attention.

About a week usually elapses from the reception of the contagion before this sore, or chancre, makes its appearance. This, if unchecked, soon produces a frightful swelling in the groin, called a *bubo*, which rapidly gives rise to abscess. All symptoms after this are called *secondary*.

Secondary symptoms generally begin after about six weeks; but the time may vary considerably. At this stage the patient becomes thin and wan; he looks dispirited; his eyes are heavy; and he complains of want of appetite, and sleep, and rheumatic pains.

Syphilitic eruptions next appear. These, in the mildest form, have the appearance of small irregular patches, of a copper-color, attended with very slight swelling; or there may be an eruption of pimples, of the size of a pea, or smaller, on the abdomen and thighs, and extending over the most of the body. As these disappear they fade out with the characteristic copper color.

Sometimes these eruptions assume a *scaly* character, on copper-colored bases. Sometimes there are large pimples full of pus, which break and give rise to small ulcers. Sometimes there are small blisters full of matter, which dry into thick scabs, under which the skin ulcerates.

Mucous patches, or *tubercles*, are soft red elevations of the skin, generally situated about the privates, where the skin is very thin; or on the mucous membrane of the mouth and throat. They are contagious, and exude a more or less copious, thin, and fetid discharge.

Syphilitic ulcerations of the throat, nose, and palate, are also very common secondary affections. They usually occur in three or four months after the infection; but this time also varies. These affections vary very much also, in severity; from a mere rawness of the mucous membrane, to terrible ulcers, which badly injure, or permanently destroy the voice, by spoiling the cartilages of the throat; eating holes through the soft palate; eroding the tonsils; and destroying the very substance of the bones of the nose, and roof of the mouth.

Ulceration of the larynx, or upper part of the windpipe, is known by huskiness, or whispering voice, suffocative cough, and expectoration of bloody matter; there is great loss of flesh, and strength, and life is sometimes lost by suffocation. All this may coincide, in point of time, with the eruptions already spoken of; all which may occupy from six to twelve months.

Syphilitic disease of bone is usually counted among *tertiary* symptoms. The bones of the face, head, and legs are most frequently attacked. It begins with tenderness of the bone, and *severe pain*, which is felt mostly in the night. This pain is shortly followed by oblong swellings of the bone, called *nodes*. If these are allowed to extend, caries, or rottenness of the bone follows; and in this way the root of the mouth, bones of the nose and eyes, and even the skull itself, may become eaten through, attended by frightful pain and discharges, and resulting in death from irritation, or protrusion of the brain, or from exhaustion and decay of the entire system.

Death may also be caused in the primary stage, by eroding, or eating chancres; which may so extensively destroy the private parts, of male or female, as to cause death by eating into the very bowels. Such a case occurred in one of the great Paris hospitals while the editor was attending there; we have a drawing of the same, made at the time, which the curious or doubtful should see.

Among tertiary symptoms must be counted those terrible eruptions on the skin, called tubercular, which may appear in from six months to twenty years, after the beginning.

They usually appear as broad. red, copper colored tubercles, or bunches, at the lower end of the wing of the nose, and on the face and forehead. They may also occur on almost any part of the body or limbs; and are very disfiguring, because they destroy the true skin, and thus leave, even when healed, indelible scars, frequently quite large. When occurring on the face, or forehead, they are terribly vexations from this cause. An experienced eye can detect, not only such scars, but all the syphilitic eruptions, of which there are many kinds, as far off as they can be seen. The *pus*, or matter from all syphilitic discharges, of long standing, has a peculiar, and often almost intolerable fetor; especially is this the case with the breath, when the throat, teeth, and bones of the palate and nose become affected.

All the more distressing and peculiar sufferings of this fearful disease are aggravated, and often in part caused by the common, old school, or mercurial treatment. In fact so similar are many of the symptoms of mercury to those of syphilis, that it is sometimes very difficult, in cases of long standing, where the system has been repeatedly saturated with that most pernicious drug, to determine accurately what part of the disease belongs to each. *Scrofula*, too, is liable to be mixed up with both.

In the great syphilitic hospitals at Paris, where we have seen several hundred such cases, almost daily for months, such distinctions are very nicely made out; but many physicians who still *salivate* their patients, know really very little of syphilis, the effects of mercury, or scrofula.

One of the most vexatious symptoms attending this disease, is the falling out of the hair of the head, and privates, which may happen in any part of the disease; but particularly during the first three months. Severe headache often attends it.

Syphilis of children is often derived from one or both parents, before the little innocents are born. A babe deriving syphilis in this way, may be born dead, or weakly and shrivelled, with hoarse voice, discharge from the nostrils, and copper-colored blotches, or ulcers, all over the body, or only about the privates. Verily is this disease hereditary, and transmissible even to the third generation.

It should be borne in mind, that persons may be afflicted with the various forms of syphilitic disease without any fault of theirs. It may be hereditary, and thus be transmitted from father or mother or both, to the third or fourth generation; thus the iniquity of parents is visited upon their children, and appears in every form which any disease can assume. We have seen whole families infected, and in no two of them were the external symptoms alike; none of whom were conscious of the nature or the origin of the disease, excepting in some cases, one of the patients, from whom all the mischief originated. The disease may also be acquired by the application of the virus to any part covered by only a thin epidermis, as the lips, the membrane lining the nasal cavities, the eyelids, or conjunctiva, to a cut, wound, or sore, to any part chafed or abraded. The infection may be received by handling or washing venereal sores; from sponges or cloths used by syphilitic patients; from bedclothes and second-hand clothing; from the seat in the privy; from food prepared by hands or fingers on which are venereal ulcers; by a kiss; and in cases of strong susceptibility, by a breath loaded with the infection, and various other ways.

The treatment of syphilis cannot be definitely laid down in a popular treatise like this. It is a disease which does not admit of safe and effectual domestic management; but often severely taxes the skill of the most experienced physicians who have seen many thousand cases. All chronic cases of eruptions and scrofulous admixtures will be very much benefited by Rush's Sarsaparilla and Iron; and this will do the more good if the disorder is one of debility. Many cures have been effected by its persevering use; and we have no hesitation in saying that the treatment of chronic Syphilis, by this medicine alone, will be found more satisfactory than by any old-fashioned medication.

In many cases, however, other remedies will be required; and a physician must be sought. To all such we would say beware into what hands you may fall. The old, or mercurial treatment is terribly destructive; and yet many of our best family physicians know of no better remedies. There is, however, a still greater danger to avoid; and that is of those mercenary impostors who occupy large spaces in the city papers, with their pretentious advertisements. Many an unfortunate has lost most precious time, and been ruined in health for life, by their ignorance, or folly. There are now no respectable physicians advertising for the cure of such complaints;— avoid them all as you should the dire infection itself.

Another "dodge" is to advertise "a report of the Howard Association," or the like, which is sent gratis. A most exhorbitant sum is then demanded for treatment; the whole thing being a sheer piece of imposition.

In regard to the best mode of treatment, we would say that long experience has taught us that the most efficacious remedies, as well as harmless, for this disease, come from the vegetable kingdom; of which we make the most ready use in the form of their active principles. In this way crude drugs and bulky medicines are quite avoided, and the stomach is never offended.

Many patients who come under our care have been quite as much injured by the use of large quantities of filthy and destructive drugs, as by any disease of a specific character. With our mild, efficient, and agreeable medicines, and a proper course of diet and regimen, we soon change the condition of these miserable unfortunates from a state of disease and despair to one of health, vigor, and hope ; giving no more medicine than is really needed, and permitting the vital powers to have full play, instead of drowning them in a sea of physic. The truth is, no real cures of disease are ever made by overpowering the system with drugs, but by the *reaction* of the organism under judicious medicinal impressions.

None should despair of being cured, as all cases are curable, except the most malignant, at the very last.

The following letter, among many others, will show how cures are very readily effected : —

LETTER PUBLISHED BY PERMISSION.

C——, MASS, May 3, 1864.

Prof. Flanders, — *Dear Sir :* — As you have requested me to state the particulars of my case, for the benefit of others, I am willing to do so, only you must not mention my name or the town where I reside. I contracted Syphilis about seven months since, as I told you when you commenced doctoring me ; and had grown worse and worse all the time, and had paid out over forty dollars, to a calomel doctor, first, who salivated me till my teeth were loose, and I only grew worse; then an Indian doctor gave me quart bottles of roots and herbs, stewed up, sweetened with molasses, but did me no good; then I bought all the medicines I could find advertised; some injured me and some did me no good; so I went on till I began to despair of ever being any better, and thought the fate awaited me as described in the Scriptures.

I had now nearly spent all my money, when one day I chanced to see a copy of your " Family Physician." I had been sick then four months; had open ulcers in my groin, and copper-colored blotches all over my limbs and body. My voice had begun to leave me, and my throat was very sore, and I was sick at the stomach nearly all the time, from the effect of the calomel and drugs of the Indian doctor and " specifics " I bought at the drug stores. I wrote to you a little over three months ago, and am now almost entirely well. I found no trouble in describing my case to you exactly, as you sent me a list of questions to answer, so that I did not forget anything.

The first medicine you sent me, by express, made me feel like a new man in less than three days from the time I began to take it, and in a week I was able to go about my work as usual. You said it took longer to cure me of the poisons I had taken than of the disease, and I believe it did. I think I sometimes, now, feel the effects of the mercury the calomel doctor gave me, in my teeth and bones; but, except that, think I am entirely cured.

Please send me a little more of your " Calomel Antidote," as you say you think you can entirely cure me of that too.

Dear doctor, I am truly grateful to you for curing me, when I was so bad. I will say that the purest remedies and medicines, from your LABORATORY, are the cheapest and the best.

Very thankfully and truly yours,　　　　W——.

N. B. — You may put this letter in your paper, if you want to, all but the last name.

Persons wishing further advice, or information, can obtain the same, free of charge, by addressing the editor of this book. For the address see page 44.

All letters are strictly confidential, and will be returned, if so requested.

The detection of Syphilis is not always easy. The eruptions on the skin are characterized by a peculiar copper color, by the absence of any pain or itching, and their assuming different forms, at the same time, in the same person. It should not be forgotten that genuine Syphilis is essentially an eruptive disease.

During the first month, an examination of the *person* will detect the syphilitic sore, or chancre ; and the linen will usually be found soiled with bloody matter.

Syphilis may be prevented by anointing the penis with sweet oil before connection, and by thorough washing and urinating, immediately after.

The female must use Rush's Preventive, as an injection, as directed at page 231.

GONORRHŒA.

This disorder is an imflammation of the urethra, arising from the contagion of an unhealthy sexual contact.

It is not merely a local affection, but if unchecked, is liable to produce a kind of rheumatism, or to cause swelled testicles, or dropsy of the testicles; also disease of the bladder and prostate gland. It may also become chronic, and converted into *a gleet*, which rarely lasts less than a year. It may also cause abscess of the glands of the groin, and abundant wart-like vegetations on the penis. Chronic cases are always liable to degenerate into stricture of the urethra.

The clap generally begins in about two or three days after exposure to contagion, by a slight fulness, and itching, at the end of the urethra; which is followed, after one or more days, by heat and swelling, extending up the course of the passage: scalding on making water; and soon after, by a more or less copious flow of greenish-yellow pus, or matter. This pus is contagious, and if it gets into the eye, will cause a similar affection there, called gonorrhœal ophthalmia, which may cause the loss of an eye, with great suffering. The eye sometimes becomes so diseased, without any known contact of the contagious matter.

One of the most disagreeable symptoms consists in painful erections, called chordee, which take place mostly in the night. The pain is often very severe, from the penis being forced into a bent shape, owing to the urethral part being so much swelled, and tender, from the inflammation. The scalding on making water, too, is often very severe, but always worse in first attacks.

The disorders, of which we have already made mention, may follow at indefinite periods, but cannot successfully be treated without large experience.

The treatment of common cases should be commenced by folding a towel, kept constantly wet with cold water, round the suffering part, which will alone greatly relieve the inflammation and pain. One end of the towel may be left dry, to cover the wet part with. It is a useless piece of folly to take those pernicious drugs, copaiva, cubebs, and the like. Instead of nauseating the stomach, and injuring the lungs in that way, procure a box of Rush's Restorer and Preventive, and put as much as you can hold on the blade of a penknife into a pint of water, and use the liquid as an injection very frequently; say every hour, or as often as possible.

If the above injection causes any irritation, put in more water, until it is hardly felt. It should be used of a strength to produce very slight irritation only, and be prepared freshly every day. A glass syringe should be used, and each injection be preceded by passing water.

After the first two or three days, the frequency of the injections may be gradually diminished.

The effect of this Restorer is truly wonderful, frequently curing obstinate cases in a very few days. Many cases, however, will occur which have been made worse by nitrate of silver injections; or by other barbarous treatment, or by neglect ; or which have become involved in some of the foregoing complications;—all which will require the utmost skill and care of an experienced physician. The same remarks will be here applicable, that we have already made at the conclusion of the last article *(Syphilis)* to which the reader is referred.

Gonorrhœa, in the female, is usually very readily cured by Rush's Restorer; it should be used every two or three hours, by putting as much as can be held on a new cent, into a pint of water, and using it very freely as an injection, with a good syringe; which should be a very large glass one, or one of the new rubber kind, which is best. Prepare the injection freshly every time.

The prevention of gonorrhœa is very easy, by means of Rush's Restorer and Preventive. It should be used as an injection, by both male and female, as previously directed, immediately after the act ; or *any time the next day*. Passing water, and thorough washing, after enjoyment, are also advisable.

The detection of this disorder is very easy, from the discharge of greenish-yellow matter; and from the characteristic stains of the same color, to be found on the linen.

No lady, or gentleman, who has once used this remedy, for discharges of any kind, will willingly be without it. It is superior to all other medicines for this purpose ; many of which are sold for five times as much.

SELF-ABUSE.

SECRET INFIRMITIES OF YOUTH AND MATURITY, ARISING FROM SOLITARY HABITS.

In approaching this subject we feel no little embarrassment, because it has been so frequently dwelt upon in catch-penny books, and so adroitly handled by worthless swindlers, that we feel reluctant to broach it. The cause of humanity, however, requires it; and plainness of speech is here unavoidable.

Both sexes, girls and boys, men and women, are the slaves, and victims of self-abuse, or solitary vice. Nor does this state of things depend a little upon a certain sham modesty, which prevents parents from warning their children against this pernicious habit. If they were early taught that it would strip the flesh from their bones, and make them weak, ugly, sick and hateful: how many might be saved from the insane asylum, or a consumptive's grave; or from a broken down and ruined constitution.

The following are some of the disorders which are caused, wholly, or in part, by this vice: — Insanity, consumption, seminal weakness, all sorts of nervous affections, decay of the spine, hysteria, nocturnal emissions wasting of the genital organs, impotence, barrenness, discharges from the urethra, bad dreams, nightmare, palpitations, fits, injury to voice, sight, or hearing, emaciation loss of memory, dizziness, diseased kidneys, and, indirectly, many other troubles.

An able writer says: —

"The patient, by neglect of himself, or from a false modesty, which is too common with this class of patients, has delayed seeking for proper medical relief, until he is completely destroyed. Body and mind are in ruins. The generative organs are so wasted as to be entirely inactive, or so diseased as to secrete but a popy, thin, and glairy fluid, having few or none of the characteristics of semen, and which continually flows away from the unconclous victim. He is finally either hurried to a premature grave by consumption, epilepsy, or apoplexy; or insanity, taking the hopeless form of dementia, has removed him from his own home to the mad-house. It is safe to say that, of all the cases of incurable insanity, a large majority are caused by involuntary seminal emissions, or by masturbation."

Dr. Woodward, Superintendent of the Hospital for the Insane, has the following remarks on this practice: — "For the last four years it has fallen to my lot to witness, examine, and mark, the progress of from ten to twenty-five cases. daily, who have been the victims of this debasing habit; and I aver that no cause whatever, which operates on the human system, prostrates all its energies — mental, moral, and physical — to an equal extent. I have seen more cases of idiocy from this cause alone than from all the other causes of insanity. If insanity and idiocy do not result, other diseases, irremediable and hopeless, follow in its train, or such a degree of imbecility marks its ravages upon body and mind, as to destroy the happiness of life, and make existence itself wretched and miserable in the extreme."

A late celebrated surgeon, in noticing the effects of self-indulgence, says: "A habit so baneful to many of our youth, that I believe it to be more destructive in its effects than a great proportion of all the diseases to which, in early life, they are liable. Were it to prove hurtful to those only whose self-indulgence gives rise to it, there would be less cause to regret the effects of it. Besides rendering the parent himself miserable, it evidently entails the severest distress upon his posterity, by generating languor, debility, and disease, instead of that strength of constitution, without which there can be no enjoyment."

Perhaps enough has now been said to show the dangers resulting from such a course; but the question of a remedy next occurs. It is not enough, when serious disorders have arisen from this habit, to leave it off, in order to obtain a cure of its results.

A suitable course of remedies, in many cases, is quite essential; and, from the great diversity of disorders which may arise, it is evident that no general directions can be given for their cure in the limits of this work. All cases attended by debility and nervous disorders, with loss of flesh and strength, will be very much benefited, and many cases entirely cured, by Rush's Sarsaparilla and Iron (see page 47.) This truly valuable medicine should be continued as long as it improves the general state of the health; but many cases arise which no single remedy will cure, To such persons we would say beware into whose hands you fall. It is a common trick with certain imposters, who call themselves clergymen, to send some wonderful receipt, free of charge. Such a person is now advertising a preparation called the Corassa compound, which is a shameless piece of imposition; and we believe it to be wholly worthless. No such plants are known to botanists. All the little books sent so freely to sufferers we believe to be upon a similar scale, and only devices to extract money from the pockets of the unfortunate. Beware of any man who sends you a prescription free of charge, made up of drugs which you can *only* obtain of him. All such are either worthless, or injurious.

The following letters published by permission, and selected from many others, best show the nature of such cases and the time required to cure them, by the aid of the most improved modes of treatment now known.

"SRINGFIELD, June 3d 1864.

"PROF. FLANDERS,—*Dear Sir:* I have been for some time afflicted, and believing from your writings that you can relieve me, I take the liberty to address you by the medium of the pen. I will proceed at once to describe my case: Nervousness, or trembling; cold feet, especially in winter; occasional pain in the right lobe of my lungs; weakness, and sometimes pain, in the broad of my back; sometimes slight pain in the spinal chord, just below the small of the back; also, occasionally in the testes, or privates; occasionally singing in my ears; sometimes dulness of hearing; tears are easily excited; eyes weak, I believe more so in summer than in winter; dimness of vision dots or specks before my eyes; gathering of films before them. My step is not so light and elastic as formerly, nor can I walk in so direct a line as formerly. The passions are more easily aroused — anger especially — and when angry, I am nervous and weak. Formerly my voice was very strong, now I think it is neither so strong nor so flexible; sometimes when reading, and frequently when talking, my voice seems to catch slightly; my delivery of speech being less rapid and easy than it once was. When I was a boy I could run eight or ten miles an hour; now, upon a little running, or other extra exertion, I am out of breath, and my heart beats rapidly. When I was a boy I rarely ever wanted a coat; even in the coldest weather, the cold did not seem to affect me; now the cold wind quickly chills me through. The grasp of my hand is not so strong as formerly. I spit a good deal of mucus. I have less inclination to female society than formerly, though naturally very fond of it. My comprehension, I think, is not so keen as it once was. My temper is not so equable as formerly, I am more easily irritated. and somewhat inclined to moroseness. My ambition for study seems diminished. I long again to revel in mental acquisitions, but my memory, once so strong and retentive, now keeps only the main points of a subject; its failure seems to rear a barrier not easily to be removed. My appetite is keen, mostly for strong diet.

"DESCRIPTION OF MY PERSON.—My height is 5½ feet, circumference of chest 33 inches, weight from 115 to 122, though persons mostly think I weigh about 135 or 140. When a boy I was muscular and robust, now I am rather thin in flesh, though my flesh is solid. Very fair complextion, though freckled and easily tanned. Blue eyes, curly hair, age 22. I am disposed to active service, though my occupation is about two-thirds of the time in doors; mostly mental labor. In nearly all kinds of medicines I require nearly or quite double the dose for common persons.

"The above evils I regard as resulting from the habit of *solitary indulgence*. I am also afflicted with nocturnal emissions; also with passing water more frequently than I should. I am, otherwise than the above, usually in good health.

"Dear doctor, give me aid as soon as possible. Please return answers well secured. Yours, respectfully, M. C."

Twelve weeks later he writes:

"DEAR SIR:—The last medicine you sent me, has nearly completed my cure. I rejoice in recovering my former health and strength of mind and body. Please send what medicine you may now think necessary to prevent a relapse.
"Yours, respectfully. M. C."

"NORTHVILLE, Jan. 10, 1804.

"*Dear Sir:* — I write to you on a delicate subject, one that I am much ashamed to own, it is the practice of self-pollution. The practice I have quit, but the disease remains. I am subject to seminal emissions, sometimes not in two weeks, then again sometimes twice a week. I have a loss of the appetite, a weakness of the limbs, pains in the groins and dulness, irruptions on the skin. I am 20 years old, tall and thin; my occupation is of a sedentary kind; I have tried one physician "private," but he done me no good, but I will trust you. Sometimes I feel that there is nothing for me to live for; I have been tempted to fill a suicide's grave. When I wrote to the physician that treated me the first time, and he promised to cure me, I had some hopes, but when the medicine done no good I began to despair; but I accidentally was reading your Physician, I thought that I could try once more; and if you cured me the blessing that I can offer shall be yours, and you will ever have my friendship. I love a beautiful and lovely girl, and had intended to get married soon, but I dare not, as I am now, to blast the hopes of another person for life, and make myself miseraable for life too. There is at most times a weeping of the urethra, which strains my limbs. I beg pardon for writing so much, and Remain yours, S."

After taking the medicines for one month he writes:

NORTHVILLE, Feb. 21, 1864.

" *Dear Sir and Friend* — For a friend you are. I write to you according to request. I received the medicine safe and sound, and commenced taking it immediately, and soon began to feel the effects of it. I feel better in mind, better in health. I grow stronger; the pains have all gone; I have not had but one seminal emission since I got the medicine; no weeping of the urethra; testicles are increasing in size and altogether much strengthened, for all which I give you credit. I will write again in about two weeks, and remain ' Yours, with respect, S."

And a month later he writes:

" NORTHVILLE, March 13, 1864.

" *Dear Sir :* — I pen these few lines to you to let you know that I am rapidly improving in health under the influence of your treatment. I little thought, three months ago, to be on this side of the grave. After I tried one physician and he failed. I gave up all hopes of ever recovering, and when I noticed your offer of advice, I thought they were all alike; and when I sent to you I thought it was money thrown away, but I have changed my mind, and now I class you as one of the benefactors of mankind. The symptoms have all left me; the pains in the back and loins have gone, my appetite is increasing, and altogether I feel a new creature; no weeping at the urethra, and no seminal emissions.

" Yours, S."

Any unfortunate person who desires further information, or advice, can obtain it, free of charge, (by addressing the editor of this book; see address at page 44). All letters are strictly confidential, and will be returned if requested.

RECEIPTS.

The following receipt was extensively sold for one dollar, and diminishes the labor of family washing nearly one half, as the best housekeepers say. It may be made by any intelligent person, as follows:—

Chromate Soap.

White Soap, Five Pounds; Washing Soda half pound; Borax, quarter pound; Soft Water 4 1-2 gallons.

Dissolve these articles in the water, about as hot as you can hold your hands in; then add of carbonate of ammonia half ounce, and let that dissolve, then set it away and let it cool; it will be nearly solid.

Mode of Using.—Put a teacupful of soap into two pailfuls of water; heat the same as hot as you can hold your hand in, and put it into the tub with the clothes, cover the tub with a thick cloth, and soak twenty minutes, then wring; then put the clothes into cold water, and add half a cupful of soap for two pailfuls of water; then boil 15 minutes, then rinse as usual. This process saves the whole labor of rubbing, unless the clothes are very dirty; in which case those that need it may be rubbed after the process in the warm water, and before the clothes are boiled.

The object of covering it with a thick cloth is to keep in the strength of the carbonate of ammonia, but is not very important. No family will willingly be without this after having once used it.

To keep Butter a Year.

Put six pounds of Turk's Island salt into two gallons of hot water; scald and skim. Put the butter in common lumps into a firkin and pour on enough of the brine to cover the butter. Keep the butter under the brine, with a plate or flat stone. This is the easiest way to keep butter perfectly sweet, and nice

Rush's Hair Mixture.

Tincture Arnica, 8 ounces; Castor Oil, 8 ounces; Tincture Cayenne, 2 ounces; Water of ammonia, 1 ounce.

Mixed together in a bottle. This is far better for the hair than any preparation now in the market. Perfume to your taste.

Bed Bug Poison.

Lard 6 ounces; Corrosive Sublimate pulverized, half an ounce. Warm the lard in a bowl, and mix with it the Corrosive Sublimate, with a wooden spoon, or flat stick: if the lard is entirely melted, it will do no harm, but must be stirred while cooling. It should be pushed into the cracks of the room, the floor, and the bedsteads. Any excess may be wiped up with paper; this will last for years, and one application is sufficient to clear the house of bed bugs; a sure remedy after all others have failed.

Best Black Ink.

Nutgalls in coarse powder 2 ounces; soft water a pint. Put them into a bottle, and let the galls remain one week; then add three-quarters of an ounce of Copperas, and after three days, half an ounce of Gum Arabic, in small pieces. This makes a beautiful glossy dark ink, far superior to the most of the ink now sold. It is not very black at first, but grows darker on using.

White Bar Soap.

Tallow or Lard, 10 lbs Sal Soda, 7 1-2 lbs. Soft Water, 3 quarts, Slacked Lime in coarse powder, 2 lbs.

Dissolve the soda in one quart of water; then boil it twenty minutes with the lime; then strain through a coarse towel double; then let it settle an hour: then strain again; then put the soda into a common wash boiler and add the tallow cold; let them remain cold a few hours, or over night; then apply a gentle heat frequently stirring, without boiling for an hour; then add two quarts of water, and boil gently one hour, with the cover of the boiler on loosely, stirring the soap up from the bottom frequently. When the soap is done it will be as thick as hasty pudding, and boil like it; it should then be poured into suitable pans, and when nearly cold cut into bars, or squares.

It will do no harm to boil the soap two hours; but it should not be boiled much the first hour.. If the grease is rancid, two or three hours boiling cleanses it.

The following receipts are mostly consistent with the laws of health; but those containing soda, saleratus, spices, and acids, are less so than the others. They are all, however, highly approved by the best housekeepers.

Eastern Brown Bread.

One quart of rye. Two quarts of Indian meal: if fresh and sweet do not scald it; if not, scald it.

Half a teacup of molasses; a little salt; one teaspoonful of saleratus; a teacup of home-brewed yeast, or half as much distillery yeast.

Make it as stiff as can be stirred with a spoon with warm water. Let it rise from night till morning. Then put it in a large deep pan, and smooth the top with the hand dipped in cold water, and let it stand awhile. Bake five or six hours. If put in late in the day, let it remain all night in the oven.

Bread of Unbolted Wheat, Graham Bread.

Three pints of warm water. One teacup of Indian meal, and one of wheat flour Three great spoonfuls of molasses, or a teacup of brown sugar. A little salt, and one teaspoonful of saleratus, dissolved in a little hot water. One cup of yeast.

Mix the above, and stir in enough unbolted wheat flour to make it as stiff as you can work with a spoon. Some put in enough to mould it to loaves. Try both. If made with home-brewed yeast, put it to rise over night. If with distillery yeast, make it in the morning, and bake when light.

In loaves the ordinary size, bake one hour and a half.

Wine Whey.

One pint of boiling milk, two wine glasses of wine: boil them a moment, take out the curd, sweeten and flavor the whey.

Arrowroot Custards for Invalids.

One tablespoonful of arrowroot, one pint of milk, one egg, one tablespoonful of sugar. Mix the arrowroot with a little of the cold milk, put the milk into a sauce pan over the fire; and when it boils stir in the arrowroot and the egg and sugar, well beat together. Let it scald, and pour into cups to cool. A little cinnamon boiled in the milk flavors it pleasantly.

Rice Waffles.

A quart of milk, a teacup of solid boiled rice, soaked three hours in half the milk, a pint and a half of wheat flour, or rice flour, three well beaten eggs. Bake in waffle irons. The rice must be settled enough when boiled.

Tomato Syrup, a delightful Beverage for the Sick.

Take the juice of ripe tomatoes, and put a pound of sugar to each quart of juice, put it in bottles, and set it away. In a few weeks it will have the appearance and flavor of pure wine; and mixed with water is a delightful drink for the sick. No liquor is needed to preserve it.

Currant, Rhaspberry, or Strawberry Whisk.

Put three gills of the juice of the fruit, to ten ounces of crushed sugar; add the juice of a lemon, and a pint and a half of cream. Whisk it till quite thick, and serve it in jelly glasses, or a glass dish.

Sago Milk.

Soak one ounce of sago in a pint of cold water, one hour, pour off the water, and add a pint and a half of new milk. Simmer it slowly, till the sago and milk are well mixed. Flavor with sugar.

Simple Barley Water.

Take two ounces and a half of pearl barley, cleanse it, and boil it ten minutes in half a pint of water. Strain out this water and add two quarts of boiling water, and boil it down to one quart. Then strain it and flavor it with slices of lemon and sugar or sugar and nutmeg. This is very acceptable to the sick in fevers.

Compound Barley Water.

Take two pints of simple barley water, a pint of hot water, two and a half ounces of sliced figs, half an once of liquorice root sliced and bruised, and two ounces and a half of raisins. Boil all down to two pints, and strain it, this is slightly aperient.

Pumpkin Pie.

Some prefer that the inside of the pumpkin be not scraped, merely removing the seeds. They are to be peeled, and cut into small strips, and stewed slowly, until thoroughly done; when done, let it steam slowly over the fire; when cold strain. A plain pie is made of equal parts of milk and pumpkin; and some prefer an egg or two to every quart of the mixture. The thicker of pumpkin the less eggs are needed. Sweeten to the taste. Some add grated lemon peel, ginger, spice, &c. These require a hot oven. It is well to scald the mixture just before pouring in to bake. The crust is not thus so apt to burn. The less of eggs the more baking is needed. Bake as soon as the pies are filled, to prevent clamminess of the under crust. The prepared pumpkin can be kept for months in cold weather, if well sweetened.

Arrowroot Transparent Jelly for the Sick.

Put a full teaspoonful of arrowroot into a basin, mixed with two of water, to make it about the consistency of starch; stirring all the time, pour it into a stew pan, and stir it until it has boiled two minutes, add a little cream, a small glass of wine, and a little sugar.

Blackberry Syrup for Cholera and Summer Complaint.

Two quarts of blackberry juice, one pound of loaf sugar, half an ounce of cinnamon, one ounce of allspice, pulverize the spice, and boil all for fifteen or twenty minutes; when cold, add a pint of cherry brandy.

Oat Meal Gruel.

Four tablespoonfuls of oatmeal, and a pinch of salt into a pint of boiling water. Strain and flavor it while warm. Or take fine oatmeal and make a thin batter with a little cold water, and put it into a pan of boiling water.

Ground Rice Gruel.

Take two tablespoonfuls of ground rice, a little salt, and mix it with milk enough for a thin batter. Stir it with a pint of boiling water, or boiling milk. Flavor with sugar and spice.

A great Favorite with Invalids.

Take one third brisk cider, and two thirds water; sweeten it, and crumb in toasted bread, or toasted crackers, and grate on nutmeg. Acid jellies will answer for this, when cider cannot be obtained.

Corn Griddle Cakes with Yeast.

Three coffee cups of Indian meal, sifted, one cup of either rye meal, or Graham flour, two tablespoonfuls of yeast, and a teaspoonful of salt; wet at night with sour milk or water, as thick as pan cakes, and in the morning add one teaspoonful of saleratus. Bake on a griddle. If Graham flour is used add a very little molasses.

Strawberry or Currant Ice Water for the Sick.

Press the juice from ripe strawberries, strain it, and put a pound of sugar to each pint of juice. Put into bottles, cork and seal it, and keep it in a cool place. When wanted, mix it with ice water for a drink. Currants can be used in the same way.

Alum Whey.

Mix an ounce of powdered alum with one pint of milk; boil it two or three minutes, then strain, and add the sugar and nutmeg to the whey. It is good in cases of hæmorrhages, and painter's colic.

Wedding Cake.

Take 4 lbs flour, 3 of butter. 3 of sugar, 4 of currants, 2 of raisins, 24 eggs. 1 ounce mace, 3 nutmegs. Bake three hours. Three lbs of frosting.

Frosting.

Whites of two eggs, half pound of loaf sugar, 1-8 ounce white starch, 1-8 ounce gum arabic; beat till it looks white and thick. Dry it in a cool oven.

Apple Tea.

Take two good mellow apples, slice them thin, pour on boiling water, let them stand an hour; pour off the water, sweeten and flavor.

Carrot and Squash Pies.

These are made in a way similar to the pumpkin pie, and are very excellent, if properly made.

Potato Pie.

Boil good mealy common potatoes, or good sweet ones, in different waters,—strain them. One part potatoes to six or eight of milk with beaten eggs, and bake with under-crust only.

Tomato Pie.

Pour boiling water over the tomatoes, letting them remain in the water a few minutes. Strip off the skins, cut in slices, sprinkle sugar over them, ane bake slowly for an hour, with upper crust.

Rhubarb, or Pie Plant Pie.

Tender stalks, with the skin stripped off, are cut into thin slices. One layer of rhubarb with one of sugar, till sufficiently thick; cover with a top crust, well closed at the edges, and prick holes. Bake slowly till done.

Potato Flour for Cakes, &c.

Take the best mealy potatoes, pare and wash well; grate them finely in a vessel containing water, until there is about as much potato as water. It is well to stir. When well settled, pour off the water carefully, dry the flour, grind or beat it, and sift. This is excellent where great lightness is an object, as in sponge cake. It may be cooked in a variety of ways for children and the sick.

Plain Bread Cake.

One or two parts of sugar, to three or four of raised dough, and one part of good cream, or milk will do. Mix all thoroughly. If too soft, add flour. Fruit can also be added. Let it raise awhile before baking.

A Plain Cake.

Two or three parts of sour cream or milk, with saleratus, to sweeten; then stir in five or six parts of flour, or meal of wheat, and bake.

Wheat Jelly.

Wheat jelly is one of the most soothing substances that can be introduced into the stomach. Soak the wheat over-night, and boil four to six hours. The jelly may be pressed through a coarse cloth or may be eaten as it is, with some kind of fruit or sauce. It is most excellent for regulating the stomach and bowels.

Apple Jelly.

Take nice fair apples, cut out the stem and eye; cut them in thin slices, without paring them, or quartering, as the chief flavor is in the peel, and the jelly part in the cores. Put them in a kettle, and put in just water enough to cover them, and boil them very soft. Then wash, and strain through a jelly bag, made of coarse flannel; put to each pint of liquid, a pint of brown sugar, add the juice and rind of a lemon cut in slices. Beat up one egg and stir it in thoroughly. Boil up 5 minutes; let it stand and cool a little : try it, if not hard enough let it boil 5 minutes longer; then skim off the scum, and pour off the clear jelly; then put it in jelly glasses.

Common Flat Jacks.

One quart of sour milk, thicken it with flour, two teaspoonfuls of saleratus, and a little salt.

Cocoa.

Boil two large spoonfuls of ground cocoa, in a quart of water, half an hour; skim off the oil, pour in three gills of milk, and boil it up again It is the best way to make it the day before it is used, as the oily substance can be more perfectly removed when the cocoa is cold.

Shells.

A heaping teacupful to a quart of boiling water. Boil them a half an hour, but two or three hours is far better. Scald milk as for coffee. It is well to soak them over night, and boil them in the same water in the morning.

Milk Lemonade.

To a half a pint of boiling water, add the juice of one lemon, and a wine glass of currant wine. Then add a pint of cold milk, and strain the whole to make it nice and clear.

An Economical Method of Making Chocolate.

Cut a cake in two small bits and put them into a pint of boiling water. In a few minutes, set it off the fire and stir it well, till the chocolate is dissolved; then boil it again gently a few minutes, pour it into a bowl, and set it in a cool place. It will keep good eight or ten days. For use, boil a teaspoonful or two in a pint of milk, with sugar.

Apple Mince Pies.

To twelve fine chopped apples, add three beaten eggs, and a half a pint of cream, (made of corn starch). Put in spice, sugar, or molasses, raisins, or currants; just as you would for meat mince pies.

Cherry Pie.

The common red cherry makes the best pie. A large deep dish is best. Add four spoonfuls of sugar, dredge a very little flour over the fruit before you lay on the upper crust.

Hot Cakes.

Scald a quart of Indian meal, with just water enough to make a thick batter. Stir in a little salt, and two tablespoonfuls of butter: pour it into a buttered pan, and bake it half an hour

Corn Muffins.

One pint of sifted meal, and half a teaspoonful of salt, two tablespoonfuls of melted butter, a teaspoonful of saleratus, in two great spoonfuls of hot water. Wet the above with sour milk, as thick as for mush or hasty pudding; and bake in buttered rings, on a buttered tin.

Orange or Lemon Syrup for the Sick.

Put a pound and a half of white sugar to each pint of juice; add some of the peel; boil ten minutes, then strain and cork it. It makes a fine beverage, and is useful to flavor pies and puddings.

Taffy.

One cup of sugar, one spoonful of butter, one of water, one of molasses; boil together twenty minutes; then pour it out into a long tin. When quite cool, cut it in squares; then let it remain until hard.

Crystalized Currants.

Take fine bunches of currants on the stalk, dip them in well beaten whites of eggs, lay them on a seive and sift white sugar over them, and set them in a warm place to dry.

Apple Snow.

Put 6 very tart apples in cold water over the fire. When soft, take away the skins and cores, and mix in a half a pint of sifted white sugar; beat the whites of five eggs to a froth; and then add them to the apples and sugar. Put it in a dessert dish and orna ment with myrtle.

Bachelor's Corn Cake.

A pint of sifted corn meal, and a little salt; two spoonfuls of butter, a quarter of a cup of cream, two eggs well beaten. Add milk till it is thin as fritter batter, bake in deep tins. Beat it well and bake with quick heat; and it rises like pound cake.

Crumpets.

A quart of warm milk, a little salt, a gill of yeast, flour enough for a stiff batter; not very stiff when light; add half a cup of melted butter, or a cup of rich cream; let it stand twenty minutes, then bake it as muffins or in cups.

Cream Griddle Cakes.

One pint of thick cream, and a pint of milk, three eggs, and a little salt. Make a batter of flour, and bake on a griddle.

Potato Biscuit for Tea.

Twelve potatoes, boiled soft and mashed fine, and a little salt. Mix the potatoes and milk, add half a tea cup of yeast and flour enough to mould them well. Then work in a cup of butter. When risen, mould them into small cakes. Then let them stand in buttered pans fifteen minutes before baking

Beef or Mutton and Potato Pie.

Take a deep dish, butter it, put in it a layer of mashed potatoes seasoned with butter, salt, and minced onions. Take slices of beef or mutton, and season them with a little pepper and salt, lay them with small bits of butter over the potatoes. Then fill the dish with alternate layers, as above described, having the upper one potatoes. Bake an hour, or an hour and a half.

Beef or Veal Stewed with Apples.

Rub a stew pan with butter, cut the meat in thin slices, and put it in with salt and apple sliced fine; some would add a little onion. Cover it tight, and stew till tender.

Boiled Apple Pudding.

Line a basin with paste tolerably thin, fill it with the apples and cover it with the paste; tie a cloth over it and boil it an hour and a half until the apples are soft.

Yankee Pudding.

Scald one pint of milk, half a pint of Indian meal, a tea cup of molasses, six sweet apples cut in small pieces; bake three hours. This is a delicious pudding.

Wheat Meal Pudding.

We may scald one part of good fresh wheat meal with about two parts of milk. Add a little sweetening, with raisins or other fruit, if desired, and bake an hour. It is better to stand awhile before baking.

An Excellent Indian Pudding without Egg.

Take seven heaping spoonfuls of Indian meal, half a teaspoonful of salt, two spoonfuls of butter, a tea cup of molasses. Pour into these a quart of milk, while boiling hot, mix well, and put it in a buttered dish. Just as you set it in the oven, stir in a teacup of cold water, which will produce the same effect as eggs. Bake three quarters of an hour, in a tin that will not spread it out thin.

Green Corn Pudding.

Ten ears of corn grated. Sweet corn is best: one pint of new milk two well beaten eggs, one tea cup of sugar. Mix the above, and bake three hours; if common corn is used more sugar is needed.

A Minute Pudding of Corn Starch.

Three heaped tablespoonfuls of corn starch, two eggs, a little salt, one pint and a half milk. Boil the milk, reserving a little to moisten the flour; stir the flour to a paste, with the reserved milk and put it into the boiling milk, add the eggs well beaten, let it boil till very thick, then pour it into a dish and serve with liquid sauce. After the milk boils, the pudding must be stirred every moment till done.

Sweet Potato Pudding

Grate half a pound of parboiled sweet potatoes, and stir to a cream six ounces of sugar and six of butter; and then add the beaten yolks of four eggs. Mix the above, and add a glass of wine, or strong cider. The last thing, put in the whites of the eggs beat to a stiff froth. Common potatoes and carrots may be used as above, only they are to be boiled soft, and put through a colander, and more sugar used.

Baked Rice Pudding.

Swell a coffee cup of rice, add a quart of new milk; half a cup of sugar, three eggs, a little salt, bake an hour and a half.

Apple Pudding.

Set your tin pail or kettle on the stove; put in a cup of water; cut in four large apples, one pint sour milk, one large teaspoonful of saleratus; mould your crust, and spread it over the top; cover it tight; bake one hour.

Boiled Indian Pudding.

One quart of milk, five gills of meal, four eggs, a teaspoonful of salt, and one of molasses; boil three hours.

ANOTHER. Three cups of Indian meal, half a cup of molasses, mixed with luke warm water, rather stiff. Boil two hours.

INDEX.

The Figures refer to pages.

www.ingramcontent.com/pod-product-compliance
Lightning Source LLC
Chambersburg PA
CBHW021524210326
41599CB00012B/1369